# THE MILLING MACHINE

## FOR HOME MACHINISTS

FUL
5

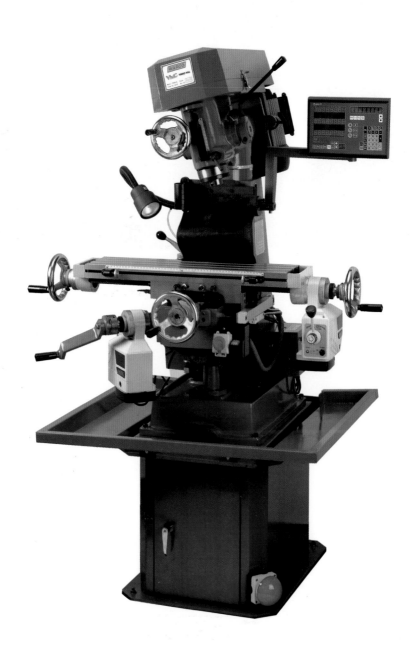

# THE MILLING MACHINE

## FOR HOME MACHINISTS

HAROLD HALL

FOX CHAPEL
PUBLISHING

Parts of this book were updated for today's American reader with regard to new techniques and tools. These updates were graciously provided by George Bulliss of *The Home Shop Machinist, Machinist's Workshop*, and *Digital Machinist* magazines.

© 2013 by Harold Hall and Fox Chapel Publishing Company, Inc., East Petersburg, PA.

First published in the United Kingdom by Special Interest Model Books, 2011.
First published in North America in 2013 by Fox Chapel Publishing, 1970 Broad Street, East Petersburg, PA 17520.

ISBN 978-1-56523-769-8

Library of Congress Cataloging-in-Publication Data


Hall, Harold, 1933-

 The milling machine for home machinists.

   pages cm

 Includes index.

 ISBN 978-1-56523-769-8

 1. Milling (Metal-work)--Amateurs' manuals.  I. Title.

 TT209.H355 2013

 671.3'5--dc23

                    2012042876

To learn more about the other great books from Fox Chapel Publishing, or to find a retailer near you, call toll-free 800-457-9112 or visit us at *www.FoxChapelPublishing.com*.

**Note to Authors:** We are always looking for talented authors to write new books. Please send a brief letter describing your idea to Acquisition Editor, 1970 Broad Street, East Petersburg, PA 17520.

Printed in China
Second printing

# CONTENTS

# PREFACE

Having provided the book 'Milling, A Complete Course', Workshop Practice Series number 35, it became apparent that many workshop owners were looking for a more theoretical book on the subject rather than the hands-on approach of that published. This book therefore attempts to fill the gap, going into much greater detail regarding the choice of machine and the accessories that can be used with it.

There are chapters covering which accessories should be considered essential from the early stages and those which can be delayed until a need arises. This includes such accessories as dividing heads, rotary tables and boring heads. Within this detail, different versions of an accessory are discussed and their plus and minus points are examined.

The book then looks at the subject of installation, although this is relatively easy compared to the requirements for most lathes.

Having covered the subject regarding which machine and accessories to obtain, the following chapters discuss the methods of holding the workpiece, and in the case of active accessories (such as dividing heads and rotary tables) give sufficient detail regarding their use for all but the most complex applications.

Having the workpiece then ready for machining, the machining operation is covered in depth. The book concludes with a simple device for sharpening the end cutting edges of a high-speed steel end mill.

# INTRODUCTION

For many years the milling machine was a rarity in the home workshop, with milling carried out on the lathe, some still using this as their only method. More cheaply available machines, though, and in a much larger range of types and sizes – particularly smaller ones – coupled I feel with greater wealth, has resulted them being much more common.

While some of this book will apply equally to both methods, it is mainly aimed at the workshop that possesses a conventional milling machine. I say 'conventional' although there are two types available: horizontal and vertical. Again, the horizontal machine is rare in the home workshop, so the book only deals with the vertical machine in depth.

Even having limited the choice to a vertical machine, there are still many decisions to be made – size, speed range, drive type, etc. All these considerations are discussed throughout the book.

While new machines are often supplied with a small number of accessories, these are rarely adequate, a major omission invariably being a cutter holding chuck. The accessories are therefore considered fully, from those that are essential, typically a cutter chuck, through to those that many will work without, such as a dividing head.

On the basis that an accessory is something that you may or may not need, then one item that is not an accessory, and one that cannot be done without, is a cutter. The range of these is vast, but mostly applicable to their use in industry. Even so, the number that may find a use in the home workshop is still quite large; which cutter to choose is, as a result, an essential part of equipping the workshop with a milling facility.

Having acquired your machine, together with the initial set of accessories, it will need installing. Fortunately this is simpler than with a lathe, but there are factors that must be taken into account; some are easily overlooked, such as whether there is enough headroom to remove the draw bar.

With the machine and all its accessories ready to be put through their paces, it is now that the real learning curve starts. The reason for this is the multitude of shapes and sizes of workpieces that will invariably surface during the machine's life. After many years using a milling machine, you will still probably find yourself looking at something quite unlike any workpiece you have attempted previously. In simple terms, the decisions to be made are: do I use a vise or an angle plate or mount the workpiece directly onto the worktable, or perhaps even produce a fixture specifically for the task? Even then, how is the part positioned suitably for machining? This aspect of the milling machine is a major part of the book's content, and for the machine owner the only way to gain the vital experience.

With the workpiece mounted and the chosen cutter in place, the next task is to start the machining operation. For this, such items as traverse direction, effect of the machine spindle not being at 90° to the machine table, speed and feed rates are covered.

The process of removing metal is not as simple as it may appear, but once the requirements are understood it will very quickly be mastered. However, choosing the method of mounting the workpiece and the order in which surfaces should be machined is a continuous learning process in view of the large number of workpiece shapes that will occur. Therefore, in terms of the time taken for machining a complete part, the process will typically be 40% deciding the method, 40% setting it up and only 20% actual machining – often even less machining time than that.

The various methods are therefore discussed in depth and illustrated by examples of actual set-ups. In a nutshell, therefore, the book works through choosing the machine and its accessories, mounting the workpiece using the chosen method of securing it, and finally machining the part. While this follows the actual workshop process, it is not easy to totally separate the use of the machine into these three sections, so do read the whole book before making any decisions regarding with what to equip the workshop.

# CHAPTER 1
# THE MILLING MACHINE

As stated in the introduction, there are two types of milling machine, horizontal and vertical, and this book will only consider the vertical type in detail as the horizontal machine is rarely found in the home workshop. However, briefly, horizontal types hold a cutter, like a very wide circular saw, on a horizontal arbor, and beyond that they work much like the vertical types. They are much less prominent in machine suppliers' catalogs and those that do feature are well outside the budget of the vast majority of small workshops.

In the case of the vertical machine, though, the range available is very wide with a large number suitable for use in the home workshop. However, should the reader have a particular wish to include horizontal milling then a universal mill having both vertical and horizontal features, **Photograph 1.1**, is a possible approach. It should be apparent from the photograph that this has both horizontal and vertical spindles.

## SIZE

When choosing a milling machine for the workshop, the main considerations will relate to the size of the workpiece that can be accommodated; this, however, is not easy as there are so many variations that fit into the equation. Of course, if you only have a very small workshop then the overall size of the machine may be the deciding factor and the tasks you attempt will then be limited by that. This is beyond what can easily be covered in this book, and it can only consider the space available for the part being machined together with the accessories used with it. These are typically a vise or angle plate and the cutter

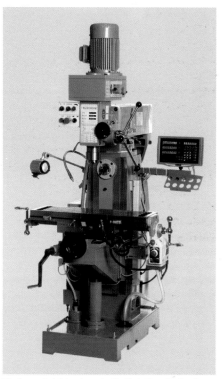

*Photograph 1.1 A universal mill with both vertical and horizontal spindles. (Chester)*

assembly fitted to the machine's spindle. The three major dimensions are, therefore, table size (length and width), cutter spindle to rear column distance, and maximum table to machine spindle height.

Advising which size to purchase is impossible, but, having decided the type of work you intend to use the machine for, do tend towards machines that would appear rather bigger than required. In the vast majority of cases a large machine

*Photograph 1.2 Do ensure that the accessories you choose are compatible with the machine. A 100mm vise on a machine with a 150mm wide table may sound suitable but, as the photograph shows, it is far too big.*

will adequately cope with small parts but obviously it will be difficult to stretch the limit of a machine that is just too small. Of course, if you are absolutely sure that you will only be using the machine for a specific interest, such as models in the very smallest scales, then perhaps you can choose a machine with very little capacity in hand.

A major factor in the decision is the size of the accessories that will be used, as it is important that, having chosen the machine and then the accessories, you do not find that the available space for the workpiece will be severely limited. The method has to be: make an initial choice of machine, then the accessories, consider the result and, if insufficient room is available for the workpiece, move up a size of machine or down a size in terms of the accessories.

This is not that easy, and you will need to know the overall size of the accessory and its fixing centers rather than just its capacity. Typically, a 100mm (4") vise on a table that is 150mm (6") wide would seem fine but, as **Photograph 1.2** shows, it is far from the case; even without the swivel base it will not

fit easily as it has no fixings allowing it to be mounted in line with the table. Also, the height of such a large vise will limit the space for the workpiece, which likewise needs to take into account the projection of the cutter chuck and cutter.

One item that needs a lot of headroom is the semi universal dividing head. If, therefore, this is to be an essential part of your workshop equipment, do take into account its height. Such considerations will be more critical if choosing one of the smaller machines.

## SPEEDS AVAILABLE

Speed range is another factor, but again the purpose for which the machine is to be used is a major consideration. The reader will no doubt know that the larger the cutter the slower it should run, and vise versa. Obviously, therefore, if the purpose of the machine is for making large items of

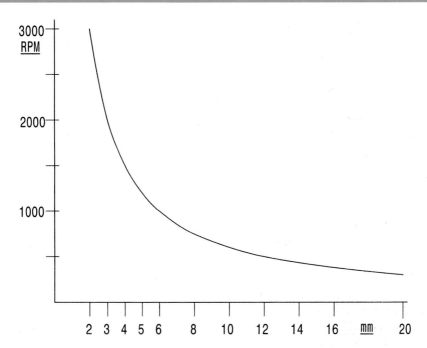

**Sketch 1.1** *Showing how cutter speed varies with cutter size for a constant cutter surface speed. Based on a 12mm cutter running at 500rpm.*

workshop equipment or models, then larger cutters, say 20mm (4") plus diameter, are likely to be used and quite slow speeds will be required. However, no matter how sizable your workpieces are, there will be occasions when very small cutters will have to be used, say 3mm (⅛") or even less. Because of this, a high speed should be available no matter how big the machine is. I would therefore aim for a speed range of at least 250 to 2000rpm. If you are purchasing a smaller machine and are unlikely to use cutters bigger than 12mm (½") then, 400–2000rpm would be acceptable.

If, however, you anticipate using larger cutter diameters, such as a fly cutter or a large slitting saw, lower speeds would be advisable – 100rpm or lower if possible. Unfortunately, you are very much in the hands of the machines that suit your other requirements and a compromise in terms of speeds available may have to be made. I recommend that you aim for a top speed of at least 1500rpm so that small cutter sizes can be used. In practice you will find that larger cutters are much more tolerant of different speeds than are the very small sizes. This is best visualized by reference to the graph, **Sketch 1.1**, that shows, with a fixed surface speed at the outer diameter, how the speed has to vary with cutter diameter.

## ELECTRONIC SPEED CONTROL

Some machines will use electronically controlled motors providing a stepless speed range, this being a considerable advantage in terms of convenience. However, excellent work has been performed for many years on machines having a very limited number of speeds available, so statements that suggest that the ability of providing exactly the correct speed is a major advantage should be given only limited credence. Avoiding the chore of belt changing is the real benefit, particularly as this can be quite difficult on some of the larger machines.

Frequently, the speed range quoted for this type of drive system is given typically as 0–2000rpm. It should be realized that at very low speeds the torque available will probably be insufficient and unable to perform any useful machining work, so it is worth making an enquiry as to the lowest useful speed. This statement needs clarification, however, as most often the torque output will be constant, being the same at low speed as at high speed.

To understand the problem it is necessary to compare it with belt or geared speed control. If we consider say a 1kw motor, and with the machine set up for top speed, then the power available will be 1kw resulting in a specific torque output. If now the mechanical drive is changed to reduce the speed to half the value, the motor will still be running at the same speed and again capable of producing 1kw of power. However, as the speed of the machine is now halved the torque available will have doubled, therefore as the speed is reduced further then the available torque will increase still further. This means that for a given sized motor there is a much greater torque available for using larger cutters and making heavier cuts.

Some machines partially overcome this situation by making machines with a combination of electronic and mechanical speed control, with the larger part of the speed range being achieved electronically but at some point down the speed range the speed is changed mechanically, this being done manually, allowing the motor to run faster than it would have done. Of course, changing the speed mechanically only

*Photograph 1.3 A typical round column mill drill. (Warco)*

needs doing if it is found that more torque is required and, as a result, the electronic control will be sufficient for most of the machining carried out.

## MACHINE VARIATIONS
Having decided that a vertical milling machine is the kind to obtain, there is still considerable variation within the type.

### MILL DRILLS, ROUND COLUMN
Probably the most common machines in today's workshops are mill drills, such as that shown in **Photograph 1.3**. These provide the X (left-right) and Y (towards-away) axes by the table's movements but the Z (up-down) axis by traversing the cutter much like a basic drilling machine (the photograph should make this clear). While the head appears very like a drilling machine, it has three important additions as well as being more robust. The vital variations are that it has: (1) a fine down feed that is calibrated; (2) a down feed

lock; and (3) a drawbar through the spindle to secure the cutter chuck. The reason for the need to secure the cutter chuck will be discussed later. I should add that the fine down feed can be disengaged and then operated as a conventional drilling machine.

Additional Z axis traverse is provided by moving the head up and down the round column. This can create a problem in some machining situations – the easiest, though least important, way of describing this is to consider it being used for drilling a large hole. First the position of the hole is started using a short center drill which is then replaced by a large much longer drill. It may be then that the head needs raising to allow the longer drill to be placed, and as the head is carried on a round column the head may rotate a little so that the alignment is lost. This can be a major problem when using the machine for milling if a change of cutter is needed and register between the already machined surfaces is lost. However, by carefully setting the height of the head for the first operation the potential problem can mostly be avoided and the problem will only surface very occasionally. This feature can on rare occasions become an advantage as the head can be swung enabling the cutter to reach the extreme areas of large workpieces.

What then are their advantages? Being economically priced is probably their main advantage, with them costing less than other types with a similar workpiece capacity. Similarly, they are bench mounted where other types of the same capacity would often be floor standing. Also, while I would not recommend using one as the workshop's only drilling machine, due to the need to break from milling to perform a drilling task, their much more robust build would be useful when needing to drill large size holes. This especially applies if the workshop drilling machine is a light duty one.

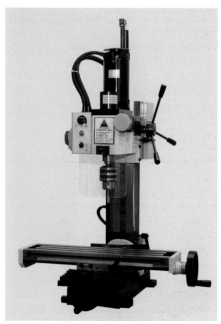

*Photograph 1.4 A small mill drill with dovetail slides for its vertical movement. (Chester)*

## MILL DRILLS, DOVETAIL SLIDE COLUMNS

These are very similar in principle to the round column mill drill, but having a rectangular column with a dovetail slide ensures that register is not lost when the head is raised. They are also made in a much wider range of sizes, with **Photograph 1.4** showing one of the smallest and **Photograph 1.5** a machine comparable with the average round column machine. For comparably sized machines they do tend to be rather more expensive, but some have the advantage of electronic speed control, or a geared head, while the round column machines are mostly belt driven.

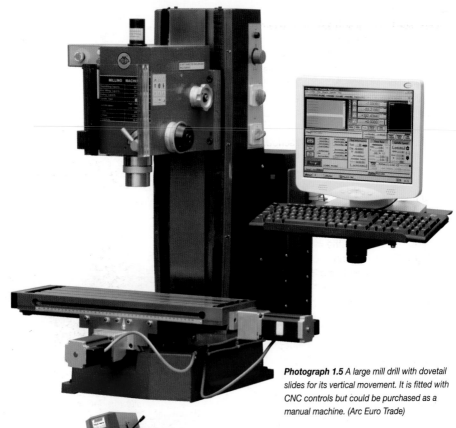

**Photograph 1.5** *A large mill drill with dovetail slides for its vertical movement. It is fitted with CNC controls but could be purchased as a manual machine. (Arc Euro Trade)*

**Photograph 1.6** *Turret mills have the facility of moving the table in all three axes but with the cutter spindle also capable of being traversed. (Warco)*

## TURRET MILLS

The only other form that is commonly available is one where all three axes (X, Y and Z) are provided with dovetail slides controlling the position of the table. The third axis (Z) is provided by a part of the machine called the knee. They do also have a head providing much the same features as a mill drill – that is, rapid down feed for drilling and fine down feed for setting the position of milling cutters (see **Photograph 1.6** for an example). Machines of this type that are available are mostly for use in industry but that in the photograph, being among the smallest available, will be about twice the price of a similar capacity mill drill; even at that price, though, speed control is likely to be by belt changing. If you have the space and the finance, then they are to be given serious consideration as their advantages make them more pleasing to use.

*Photograph 1.7 Machines can be obtained with either R8 or Morse taper sockets in their spindles. The photograph shows collets for these: R8 top, Morse bottom. As can be seen, they are quite different.*

## SPINDLE BORE TAPER

Only two types of spindle bore are commonly available, Morse and R8. As an example, collets that fit these are shown in **Photograph 1.7** – R8 top, Morse below. Morse is a shallow angle taper frequently used in home workshop machines and has a range of sizes, but only two, numbers 2 and 3, are common.

The R8 taper is a single size taper and is used in the very popular industrial Bridgeport milling machine but is now finding its way into smaller machines. Collets are made for this form of taper allowing cutters to be held without the need for a cutter chuck. A major benefit of this method is greater rigidity as it avoids the overhang of the chuck. Morse taper collets are also available that would provide the same benefits but seem to be less popular (no doubt due to the shallow taper which results in the draw bar needing a substantial hit to free the collet from its taper socket). With the R8 having a much wider angle, as can be seen in the photograph, the collet releases itself much more readily. There will be more about Morse and R8 collets later in the book.

The R8 taper will only be available on the larger mill drills and turret mills and then frequently as an option, Morse or R8. In this case, my recommendation would be to choose the R8 machine.

*Photograph 1.8 Some machines have heads that can be tilted.*

## PIVOTED HEADS

A few machines provide the facility to rotate the head, clockwise and anticlockwise, as viewed from the front, enabling the cutter or drill to approach the worktable at an angle (see **Photograph 1.8**). Do not be fooled into thinking that this will enable sloping surfaces to be easily machined; it will not unless the surface is narrower than the width of the cutter being used. Of course, if you use a fly cutter then wider areas can be surfaced, but their main use will be drilling holes at an angle.

Personally, I would attempt to find another method as getting the head accurately back into position for normal milling can be quite difficult. This will be covered in Chapter 7. Rather than pivot the head, some very small machines tilt the column at its base to achieve the same result, that in **Photograph 1.4** being an example. I do not see the facility as a major benefit and it should be accepted as just being there if all other aspects of the machine suit your requirements.

## TABLE STOPS

Adjustable stops that limit the movement of the table left and right (X axis) are omitted from some small machines, which is an unfortunate omission as they can be very useful, typically when milling a closed end slot. If two machines meet all your other requirements but only one is fitted with these stops, then go for this machine even if it is a little more expensive.

While less useful, Y axis stops are also worth having, but with the exception of some industrial machines I am not aware of them being on any home workshop caliber machine. They could be a candidate for a shop-made task once the machine is up and running.

Z axis stops are useful, mainly when using the machine for drilling, and are common on most, if not all, machines, so when selecting your machine for purchase it would be worth checking but I do not rate them as absolutely essential.

## DRAW BAR

This is a standard part of a vertical milling machine; its purpose is to secure items mounted into the spindle's taper so that they cannot become free when in use (more about this in Chapter 3). To achieve this, the bar passes through the full length of the spindle from the top and is threaded into the taper of the accessory, typically a cutter chuck. If the machine has an R8 taper, then the draw bar will have a $\frac{7}{16}$" in x 20 UNF thread, but in the case of Morse tapers the threads vary and this can result in more than one bar being required to suit the available accessories.

In the case of a first machine, if you purchase your machine from one supplier and your accessories from another, you will need to take account of this potential problem, as you will if you purchase additional accessories at a later date.

## POWER TRAVERSE

This is only available as standard on a very limited number of home workshop machines and then only for the X axis. Turret mills can, however, have power feed added to both the X and Y movements. Even then the facility is invariably an optional extra rather than fitted as standard and can be purchased with the machine, or at a later date for the workshop owner to fit. In the case of the popular mill drills some suppliers provide them as an optional extra, quoting the size of machine that one will fit, but as other makes are very similar they may also fit other machines, perhaps needing adapting in some simple way. Before committing yourself to purchasing one, do check very carefully to be sure that the unit will adapt to your machine. Power traverse for the Z axis is only available on industrial caliber machines.

Even for the X axis power traverse is far from essential for most, but if the machine is to be made considerable use of then I would recommend it being fitted. The feed rate is adjustable and can be set lower than the operator would probably find acceptable if traversed manually. This enables a superior finish to be achieved.

## MICROMETER DIALS

These will be fitted on the feed mechanisms for the three axes, and ideally each should be able to be zeroed as this considerably minimizes the amount of calculation required when setting on an additional depth of cut. This facility may be missing from some of the smaller economy machines. Do give this serious consideration as a fixed dial will make the machine much less easy to use.

## DIGITAL READOUTS

These are additional to the micrometer dials but are rarely fitted as standard and even then not always on the three axes. Read-out kits are readily available, so adding them at a later date is a possibility, but as there is considerable variation between machines the chosen method of fitting them will be the responsibility of the workshop owner.

In addition to being able to be zeroed at any point, a major advantage is the ability to display both imperial and metric dimensions. If say, you have an imperial machine and wish to make items having metric dimensions then this is a major plus. Like power traverse to the X axis, if you intend to use your machine a great deal then they are a very worthwhile addition. Of course, if you can justify the expense then they are worth having even if your use of the machine will be limited.

## METRIC/IMPERIAL

Choosing which measurement system your machine should be calibrated for is very much the reader's choice. My own personal opinion is, however, that the metric system will eventually be the standard – for example, the large US firm I worked for had started to use the metric system for its products sold worldwide. Because of this, I would suggest that the reader seriously considers that his or her workshop is equipped with metric machines.

# CHAPTER 2
# THE CUTTERS

Excluding those cutters that perform a very specific task (a dovetail cutter, for example), the choice of cutter to obtain is often dependent on the machine in which it is to be used. The reason for this is that HSS (high speed steel) and carbide cutters have quite different characteristics and some are more at home on one machine than on another.

Let us start with the more recent carbide tools. To gain the most from these, they need to be run faster and at a higher feed rate than HSS tools. This makes them ideal for use in industry because there is obviously a time saving. However, they are not produced with such a sharp edge and, having a lower cutting angle, sometimes negative, they place a much greater load onto the machine using them. This, coupled with the higher speed and feed rate, calls for a more robust and powerful machine, making them unsuitable for use in this way for everyday tasks in the average home workshop (unless the owner is fortunate enough to own a larger machine than normal, by which I mean something rather more robust than the largest mill drills). They can of course be run at a lower speed and feed rate, minimizing their benefits. Even if this is attempted they will still need a robust machine, at least the largest mill drills. HSS end mills are produced, however, with a much steeper helix angle and a sharper edge, and as a result place much less load on the machine, making them ideal for the lighter weight home workshop machines. They also have four cutting edges rather than one or two in the case of the smaller tipped cutters.

**Photograph 2.1** compares two types of tipped tool with a large (20mm) HSS end mill and should illustrate the differences, and why the lower rake and less sharp cutting

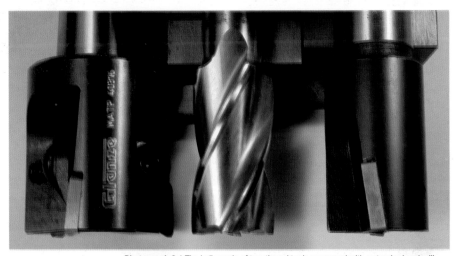

**Photograph 2.1** The helix angle of two tipped tools compared with a standard end mill. Note that the cutter on the left has a very slight angle and that on the right is even negative.

*Photograph 2.2 A tipped tool that attempts to emulate the cutting edge of a standard end mill.*

edge has the effects that I mention. It can be seen that while the cutter on the left has a slightly positive rake, that on the right is even negative, as a result increasing the load on the machine substantially.

The smaller tipped tool seen in **Photograph 2.2** is a more recent development, and it uses a specially shaped tip that seeks to emulate the characteristics of the HSS cutter. Even so, the cutting edge will be less sharp and will therefore still place more load onto the machine, though less so than the others. Also, it only has two cutting edges compared to four on a comparable sized HSS end mill, and therefore if the feed per tooth is to be kept the same the rate of feed will have to be halved.

The advantage of the smaller tipped cutter is not one compared to the HSS version but over the abilities of the two larger tools. These have tips with straight edges which means that as their edge is at an angle to the cutter's axis it will produce a curved face if used to mill a step. If you are unsure of this, place a rule onto the edge of a piece of round bar but at an angle and you will see that the rule is on a greater radius on either side of where it touches the bar, therefore it would cut a convex surface.

The larger cutters are therefore only suitable for surfacing, while the smaller with its specially shaped tip will perform the same functions as a standard end mill. A word of warning here – some small tipped end mills have tips with straight sides and in line with the cutter's axis; because of this they are able to produce a straight sided step but will demand very much more from your machine than a standard helical end mill.

There is one other consideration, and that is that carbide tools are much more suited for machining iron castings as these occasionally have very hard spots on their surface which will very easily destroy the cutting edge of an HSS tool. Because of this, one may have to be obtained just for that purpose, and if being used on a smaller machine then a slower feed rate than normal will have to be used. Depth of cut should also be only just deep enough to get below the surface skin.

## WHAT THEREFORE SHOULD BE OBTAINED?

There is another factor regarding carbide tipped tools and that is that they are not available in the smaller sizes. Because of this, no matter what size of machine you are using, smaller end mills will have to be HSS, or solid carbide. I have not mentioned solid carbide tools before but these are almost identical to the HSS tooling, but again will not have quite such a sharp edge and are much more expensive.

For larger cutters then, tipped tools can be considered but they are much more expensive; even a budget one will cost as much as many HSS cutters of the same size and a set of tips as much as one. Because of this I would suggest staying with HSS end mills reserving tipped tools just for machining the surface of cast iron casting and other difficult materials (incidentally, I refer to machining the surface of castings – below this the material will machine as easily as mild steel and HSS tools are perfectly adequate for the task).

There is another factor that I have not discussed and that is whether the cutter's shanks should be plain (maybe with a flat) or have threaded ends – the advantages of the two types are covered in Chapter 3.

## TO RECAP

Briefly, you will have to use HSS end mills for the smaller sizes, whatever caliber of machine you have, as they are the only ones available. Above this, still use HSS end mills for the average mill drill and smaller, but for the very largest round column mill drills and above a tipped end mill with the helical tips could be considered if the cost can be justified.

Larger diameter tipped tools are mostly only suitable for surfacing, but if used for the task purchase the smallest diameter available for the very smaller machines and preferably ones having a positive rake – compare the two seen in **Photograph 2.1**.

## END MILLS

Standard end mills can be used for surfacing and machining steps and because of this there is no precise link between the size of the cutter used and the dimensions of the component being made. For this, therefore, I would advise that you purchase the largest size that your cutter chuck will hold, that is typically 14mm for a 12mm chuck and 20mm for a 16mm chuck, and use them at their maximum wherever possible. Typically, that would be 8mm wide for a 20mm cutter, this being covered in detail in Chapter 16. Of course, if you only wish to cut a 3mm wide step with a 20mm cutter that is perfectly acceptable, and preferable to using a smaller cutter as it is more robust. Imperial sizes equivalent to the 12 and 20mm end mills would be ½" and ¾".

At this size I would advise having three in the workshop: a new one to be kept in reserve, a relatively new one for machining mild steel and similar, and a well-used one for machining cast iron. When those in use become too blunt, the new one can be brought into use and the bluntest kept for sharpening with others at a convenient time. As will be seen later in the book, sharpening the end cutting edges is very easy and well worth doing.

End mills will occasionally be called upon to produce a channel through a part and this is best done with a cutter equal to the width of the required channel – that is, typically, a 6mm cutter for a 6mm channel. Just how many smaller cutters can only be decided by the reader taking into account the intended purpose of the workshop.

For these smaller sizes I would suggest two of each chosen size, one in reserve, and one in use. When the latter becomes too blunt, then place it in the 'to be sharpened later' box and start using the new one.

## SLOT DRILLS

Standard end mills cannot be plunged into a workpiece for cutting enclosed slots or recesses as they are not center cutting, as **Photograph 2.3** shows. Because of this, slot drills have to be used and should ideally be available in the workshop for each slot size required. As slots are frequently to allow

**Photograph 2.3** *A standard end mill cannot be plunged as it is not center cutting.*

**Photograph 2.4** *Left to right: slot drill, standard end mill and center cutting end mill.*

adjustment for some mechanism with a screw passing through it, cutters of 5, 6 and 8mm diameter would suit M5, M6 and M8 screws; the workshop owner will of course have to make the choice depending on the screw size most commonly used and whether the shop works in metric or imperial. They do not normally get much use, so one of each chosen size should be adequate.

However, end mills are now available that are center cutting, avoiding the need to use the slower slot drills. At the time I write this, they are only available from a limited number of companies so you may have to search for a supplier. A disadvantage of these is that they cannot be easily sharpened in the home workshop without losing their center cutting feature (see Chapter 17). **Photograph 2.4** shows the three types: slot drill, common end mill and center cutting end mill.

### VERY SMALL END MILLS
These are usually very small dual-purpose mills that are suitable for both end milling and slot milling tasks, and they are frequently known as 'throwaway' or 'mini' mills. The difference is that they have three cutting edges rather than the two or four that the others have. One edge passes center, permitting it to plunge, enabling closed end slots to be machined. Commonly, these

have a 6mm shank and cutter diameters of 1–6mm, with a similar range in imperial sizes. Almost certainly at some time cutters this small will be required at least in the larger sizes 3, 4, 5 and 6mm (⅛" to ¼"), so it is a question of whether they are purchased initially or individually as a need arises.

As an alternative to the mini mills, conventional end mills and slot drills are available at these sizes, but they are not dual-purpose and are more expensive. My suggestion therefore is to obtain mini mills in the chosen sizes. Incidentally, at the smaller sizes you will need a milling machine with a very high speed, at least 1500rpm but preferably higher.

### STANDARD OR LONG LENGTH
All the above can be had in two lengths, standard and long. Standard lengths should be acceptable for almost all requirements and are of course more rigid, so I would advise obtaining longer cutters only when a need arises.

### OTHER HSS CUTTER FORMS
There are numerous other cutter forms but only a few are occasionally found in the home

**Photograph 2.5** *Left to right: ball nose end mill, rounding cutter, dovetail cutter, T-slot cutter and shop-made T-slot cutter.*

workshop, none of which should be obtained until a definite use is foreseen.

A ball nose end mill will cut slots with a round bottom or machine an internal fillet between two flat surfaces at 90° to one another. They are available in all end mill sizes including the mini mills.

Rounding end mills are available that place an external radius between two flat surfaces at 90° to each other. A separate cutter will be needed for each radius required with sizes up to 4mm with a 12mm shank, and 6mm with a 16mm shank. Standard imperial sizes can also be had.

Dovetail cutters for machine slides and T slot cutters for machine tables are also available in all the standard sizes but unfortunately some designs for shop-made tooling call for non-standard T slot sizes and a suitable cutter will have to be made in the workshop itself. However, a woodruff cutter, used normally for cutting slots for woodruff

keys, (used to provide the drive between a spindle and gear) could be considered.

The above are available with standard end mill shanks, threaded or plain. **Photograph 2.5** shows examples; a shop-made T slot cutter is on the right.

## SLITTING SAWS

Slitting saws will occasionally find a use in the home workshop, though often the task can be carried out manually using a hack saw, but the result will often be visually less acceptable. A common task is to make a cut into a round hole so that the hole can be closed to enable a close fit to be achieved or even locked completely.

These saws are available in a very wide range of outer diameters, widths, bores and tooth spacing; **Photograph 2.6** shows some examples. Bores will commonly be in the range of 5–25mm, outer diameter 20–100mm and

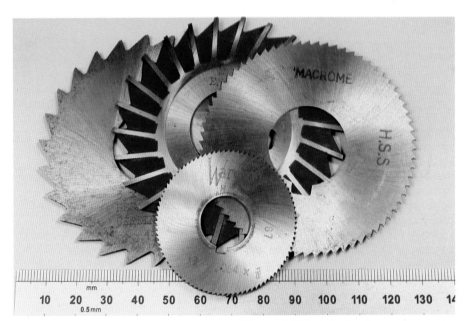

widths 0.5–5mm. Of course, imperial sizes are also available, with ½" and 1" bore sizes being most common. Because of this wide range I would suggest that purchase of these should be delayed until a need for one occurs, and even at this stage would take into consideration the following: for shallow cuts, tend towards smaller outer diameters, particularly if a thin saw is being used; if a deep cut is needed, use a large tooth size and the widest that the project will permit, as thinner saws can often wander.

**Photograph 2.6** A selection from the very wide range of slitting saws available.

**Photograph 2.7** A set of fly cutters, before being sharpened.

## FLY CUTTERS

These are single point cutting tools that have been very common in the home workshop over many years, particularly for machining large surfaces, frequently being done on the lathe. Availability of more economical milling cutters have now reduced their use. Their advantage compared to a large diameter tipped cutter is that they can be sharpened to repeat the cutting edge of a HSS end mill and will therefore place less load onto the machine in which it

is being used. They are also much less costly. **Photograph 2.7** shows a set of commercial fly cutters ( note that at the time the photograph was taken the tool bits had not been ground to the shape required). If you anticipate machining large surfaces, then a fly cutter with HSS tool bits would be worth considering.

Using all the above cutters is covered in Chapter 16, which should be read in conjunction with this chapter before coming to any conclusions regarding which cutters to obtain.

# CHAPTER 3
# CUTTER HOLDING DEVICES

Having purchased your milling machine, often it will have been supplied with a drill chuck. With this available it would then be easy for the newcomer to milling to conclude that the drill chuck will be adequate for holding an end mill when used. This, however, is very far from the case! The reason is that the helical form of the end mill will be attempting, as it rotates, to draw the cutter from the chuck, something that it will inevitably do, resulting in the machined surface becoming lower as the workpiece is traversed. This is due to the grip of the drill chuck being inadequate, and a more secure method must be used.

## SCREWED SHANK END MILLS

The collet chucks that are used with screwed shank end mills are preferred by many, myself included, as it is impossible for the cutter to be withdrawn even if the chuck is only lightly tightened. With this chuck the collet has in its base a thread into which the cutter is threaded so that the cutter's end is firmly against the base of the chuck's body. If then, under load, the cutter turns a little, it will thread further into the collet pulling the collet more firmly into the chuck's taper while keeping the end of the cutter firmly against the base of the chuck; as a result, the cutter will remain axially in the same position. **Sketch 3.1** should make the principle clear. I should add that there is some means of preventing the collet from turning, the method depending on the design. However, this system appears to be somewhat regional and I understand it is not common in the United States.

Even now with a foolproof chuck the helical cutter will also be attempting to withdraw the chuck from the taper in the machine's spindle and, with the cut being intermittent, vibration will be set up that will assist the process. These two factors can result in the chuck falling from the spindle with possibly serious consequences. This is the reason for the draw bar, which must be used in all cases

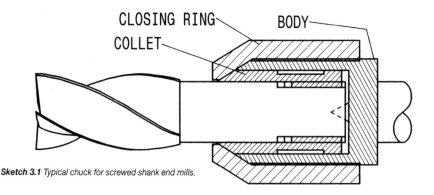

CLOSING RING · COLLET · BODY

*Sketch 3.1 Typical chuck for screwed shank end mills.*

**Photograph 3.1** *A cutter chuck suitable for both plain and threaded shank end mills. While I do not have a photograph available, I would recommend a chuck for holding threaded shanks.*

where milling is taking place even when non-helical cutters are making the cut.

When using the machine for drilling, however, the drill is attempting to force the chuck even more firmly into the taper and the problem does not exist, except maybe at the point that the drill breaks through. Because of this the tapers fitted to drill chucks are not normally provided with a thread for the draw bar, R8 tapers excepted.

## PLAIN SHANK END MILLS

Collet developments, and price reductions, have to a small extent in recent years minimized the benefits of the screwed shank system, as collets that provide a more substantial grip are now readily available. The most common is the ER type which is available in a range of sizes and when being considered for holding milling cutters ER20 will take 12mm (½") shanks while both ER25 and ER32 take 16mm (⅝") shanks. **Photograph 3.1** shows an ER32 collet together with a chuck and shank.

While an ER series collet is capable of a substantial grip, it does need to be very firmly tightened, and if this is not done then the cutter can still withdraw itself. However, with the screwed shank chucks even hand tight will work, but a little tighter is to be preferred; one manufacturer gives them the name 'Autolock', meaning the grip on the cutter will occur automatically once a cut is started. The other common collet, the 5C, **Photograph 3.2**, is most certainly not suitable as its internal angle is too large.

**Photograph 3.2** *A 5C collet. These are not suitable for holding end mills, as the internal angle is too great to achieve an adequate grip.*

An alternative method is to use collets that go directly into the machine's taper, these being available for both Morse and R8 tapers (see **Photograph 1.7**). Both have the advantage of there being no chuck to hold them and therefore no overhang from the machine's spindle. This results in a more rigid cutter. Also, it increases the maximum cutter-to-table dimension that just occasionally may enable a part to be machined that would otherwise be too large.

Both types need access to the draw bar to secure the cutter and the very narrow angle of the Morse will necessitate a blow with a soft hammer on the top of the draw bar to release it. The R8, having a greater angle, is largely self-releasing and will only need a slight tap at the most. Comparing the R8 with the 5C, the reader may question my statement above about the 5C's angle and ask why the R8 is OK, as it appears very similar. It is true that there is little difference – 20° internal for the 5C compared to 16.50° for the R8. The smaller angle of the R8 does of course make an appreciable difference, but so does the draw bar as this will have less friction compared to a closing ring, which the 5C would require.

If you have a machine with an R8 spindle taper, then using the R8 collets would obviously be your first consideration. However, if you prefer a chuck to take threaded shank cutters, these are available with solid R8 shanks and would be my first choice.

With the Morse taper being so shallow, then adequately securing the cutter will easily be possible, but you may consider that firmly hitting the end of the draw bar repeatedly to be an inconvenience and a force you would prefer not to submit your machine's bearings to. I personally would recommend using a chuck for threaded shank cutters.

## CUTTER SIZE

Another factor is the size of cutter that the chosen system will take, though I believe only a machine with a number 2 Morse taper may present a problem. In this case, if you are wishing to use Morse taper collets you will be limited to 12mm (½") shank diameters (14mm cutters). Should you wish to use larger cutters, then you will have to choose from the other systems available, all of which can cope with 16mm (⅝") shank diameters (20mm cutters); in the case of ER collets, this is size 25 or larger.

R8 collets are available up to 20mm (¾") diameter, as are ER32 collets, but as the next size shank is 25mm then these are limited to cutters with a 16mm shank size.

## TAPER TO CUTTER ADAPTORS

Adaptors are available that fit into the taper and are bored to take a single size shank which is then secured using a grub screw; for these, cutters with flats on their shanks are required. If purchasing one for each size of shank, then it would be less costly to go for a complete collet chuck, but if one is required for a size for which you do not have a collet it would be a viable option. You could of course consider making them in the workshop as they are relatively simple items, only concentricity of the bore with the taper being important.

*Photograph 3.3* Slitting saws and gear cutters come in a very wide range of bore diameters, so more than one arbor will probably be required.

## LIMITATIONS

Do take note that if you have a variety of cutter types then a chuck intended for use with threaded shank cutters must not be used with plain shanks, as the grip will not be sufficiently secure. Otherwise, Morse, R8 and ER collets can be used with both forms; the adaptors will, however, require shanks with a flat which could be ground on if absent.

Having said that collets used for threaded shank cutters must not be used for plain shanks, they can of course be used for holding such items as edge finders, as a substantial grip is not required by these.

## ARBORS FOR DISK TYPE CUTTERS

Most workshops will find a use for slitting saws at some time, so an arbor to support these is essential. I say 'an arbor' but the saws are made with a variety of bore sizes so you may require more than one. Some commercial items are fitted with adaptors to provide for more than one bore size.

**Photograph 3.3** shows two with saws fitted: the larger has a taper shank while the smaller has a shank the equivalent to a threaded end mill. Obviously, as different sizes may be required it will be a case of make or purchase as the need arises. Another type of cutter that requires an arbor is a gear cutter, and these are also made with differing bore diameters.

# CHAPTER 4

# METHODS FOR SECURING THE WORKPIECE

Securing the workpiece for machining is a complex subject due to there being a very wide range of possible workpiece shapes. Even so, they mostly fall into one of three basic methods: (1) using a vise; (2) using an angle plate; and (3) mounting directly onto the machine table. There are of course special methods that will surface occasionally, such as using some specialized item to hold the workpiece (a dividing head, for example). I should add here that this chapter is about choosing the required accessories, and putting them to use will be covered later in the book.

## VISES

Selecting a suitable vise is not easy due to the very large number of types and sizes that are available, many not being compatible with the size of the small home workshop machine. As well as being too large in terms of overall dimensions, they are more robust than the majority of workshops will require. This I believe is due to the designs originating from those used on the horizontal machine which can place greater loads on the workpiece.

**Photograph 4.1** *Two milling vises compared: a 100mm vise and a 50mm vise. The larger is too big for all but the very largest home workshop machines.*

Before attempting to choose a vise, it is necessary to define what additional features it has to include. By this I mean swiveling, tilting, a combination of these, or is it to be just fixed? My advice would be to avoid purchasing a combination vise as it would be inconvenient for day-to-day use and less robust.

## MILLING VISES

**Photograph 4.1** shows a typical large milling vise with a swiveling base, and I question the usefulness of this feature. However, in some cases the base can be removed and the vise fitted directly onto the worktable. Even in this case, though, some do not provide fixing points for it to be mounted along the table, that in the photograph being typical.

Most vises of this form are quite large relative to their holding capacity, so do check carefully that the one being considered is compatible with the machine's table size and tee slot positions (see also **Photograph 1.2** and associated text). Having mentioned capacity, most of these vises open up to a dimension close to the jaws' width. Another important factor when considering this type of vise is their weight, as the larger ones can be very heavy and may be more than the workshop owner would like to move around.

An interesting feature of the vise in the photograph is that both jaw faces can be removed and replaced on the outer ends of the jaws, enabling larger parts to be gripped when resting on the top of the jaws. However, I consider this will be of limited use and certainly not a reason for obtaining one of this type.

While I feel the swiveling base will get very little use, they are more readily available in smaller sizes such as 50mm (2") and 75mm (3") jaw widths, whereas for most fixed base vises the minimum is 100mm (4"). I am aware that smaller sizes are now appearing

*Photograph 4.2 A toolmaker's vise.*

in some suppliers' catalogs, no doubt due to the increase in the number of small milling machines now being used, the smaller vise in the photograph being an example. This has 50mm wide jaws, and a 75mm version is also available.

## TOOLMAKER'S VISES

Another vise type worth considering, especially for the smaller machine, is a toolmaker's vise, typically as seen in **Photograph 4.2**. A significant feature of these is that their method of being fixed to the table makes them easily adaptable for differing tee slot spacings and table sizes. To illustrate this, **Photograph 4.3** shows one on a Myford cross slide, and **Photograph 4.4** the same vise on a largish milling machine. Another useful feature is that all the external surfaces are accurately machined relative to its jaws, making it easy to position accurately on the worktable or use on its side or end.

The securing clamps are not supplied with the vise and will therefore have to be shop-made, which is perhaps unfortunate, but it does increase the vise's adaptability as the clamps can be made to suit differing machines, as the photographs show.

Photographs *(above)* **4.3 and 4.4** *(left) The method of mounting for a toolmaker's vise permits them to be used on a range of table sizes. The photographs show the same vise mounted onto a lathe cross slide and a large mill drill.*

Photograph **4.5** *This budget tilting vise could be considered if only small items need to be held.*

## SPLIT VISES

These are vises where the fixed and moving jaws are separate items that are individually secured to the worktable. Their main advantage is that they can secure a very wide range of workpiece sizes. I do not have an example of these to show, but the term 'split vise' accurately describes them, as both parts are very similar to the ends of a conventional vise.

## TILTING VISES

Earlier, I advised against obtaining a tilting vise as your main vise, but the need may surface very occasionally and I would suggest obtaining one of the economy versions that are readily available (see **Photograph 4.5**). These are quite robust little vises and are well worth considering providing they are adequate in terms of the size of workpiece that can be held. If of course you expect to use a vise frequently for inclined work and are looking for a more substantial one, or with a larger capacity, then do consider those based on a conventional milling vise but purchase this additionally to your main vise.

## DRILLING VISES

If the milling you intend to undertake most definitely falls into the light duty category, such as most seen in this book, then just possibly a drilling vise may be OK, but only in the case of the better ones, that in **Photograph 4.6** being an example. What then should be looked for? Acceptably accurate is the obvious criterion but the near absence of jaw lift is another and of equal, if not more, importance. Accuracy is obvious but the reader may not be aware of jaw lift which occurs when a part is being held by just the top of the jaws, **Sketch 4.1** illustrates this.

Genuine milling vises attempt to overcome this by making the assembly close fitting and with a long jaw and keep plate (the plate below the bed of the vise). With a reasonably adequate drilling vise, it is frequently possible to easily increase the length of the keep plate to improve the situation appreciably; the extension must, however, be as long as is possible and in front of the jaw to gain the most benefit.

If, like me, you would like a minor challenge, then modifying an economy drilling vise by machining all important surfaces to improve accuracy and fitting a much longer keep plate can produce excellent results (see **Photograph 4.7**). Incidentally, this has a larger capacity than that in **Photograph 4.1** which is far too big for my machine table. Having added the grooves at the side it also

**Photograph 4.6** This drilling vise is well made and could be considered for use as a milling vise where the work being undertaken is relatively light duty.

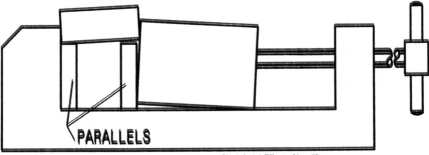

PARALLELS

**Sketch 4.1** Effect of jaw lift.

benefits from a toolmaker's vise type fixing. I should add that this is not the same vise as in **Photographs 4.3 and 4.4**; they are, though, smaller but similar drilling machine vises that have been suitably improved.

## OTHER TYPES

Some suppliers have a very wide range of vise types available, making it impossible to comment on them all individually. Many are only slight variations of the above, but many are quite different. Do, therefore, before obtaining the vise for your machine, spend time researching all that are available. Unfortunately, no one supplier can stock them all so it will be a major task. In doing this, ensure that the size of the vise is both adequate for your needs and compatible with the size of your machine.

I would have liked to have presented a more definite conclusion with regard to

which to choose, but the range in terms of design and size is far too large, coupled also with a very wide price range. For the reader, therefore, it is a much greater task researching the subject than for any other accessory, perhaps even for the machine itself.

## ANGLE PLATES

These are, in my estimation, a much under-used method of supporting a part for machining, a subject that will be aired later in Chapter 10, but fortunately choosing one is a simple task, as apart from size there are only a few variations.

*Photograph 4.10* A typical tilting angle plate.

*Photographs 4.8 (top) and 4.9* Types of fixed angle plates compared.

Choosing the size is obviously dependent on the size of the machine table and I would suggest a starting point of just shorter than the width of the table – typically, if the table width is 150mm then a minimum of 100mm. Having decided on a size, there are two versions to choose from, slotted and plain (see **Photograph 4.8**); the latter is only useful for special tasks, and the slotted plate is definitely preferred. Finally, the choice has to be made between an open ended plate or having webs like those in **Photograph 4.9**.

In the case of a small plate for use on an average mill drill or smaller, then an open ended plate has advantages. Access to the clamping nuts on the rear will be easier and clamp bars fitted behind and in front of the angle plate can project beyond its edge, which in some cases can make an impossible set-up possible. When larger angle plates are being chosen for large machines, then plates with webs will provide the extra strength that will be beneficial when a heavy cut is being taken.

If the budget will stretch to it, two identical plates can be very useful when attempting complex set-ups.

## TILTING ANGLE PLATES

These are far from essential in most workshops and should only be acquired when a definite need is foreseen, but do not purchase one in place of a fixed angle plate, attempting to make the one suffice for the two applications. Even so, the one in **Photograph 4.10**, having tee slots rather than open slots, may for the occasional

*Photograph 4.11 An example of the widely available clamping kits, probably too large for most home workshop use.*

set-up have advantages over a more conventional fixed angle plate with open slots, but I recommend it as additional to the shop's fixed angle plate rather than in place of it.

## MOUNTING DIRECTLY ON THE MACHINE TABLE

Mounting workpieces directly onto the machine table is the simplest of the three methods, but beyond that has other advantages. Briefly, these are that it avoids any inaccuracy in the vise or angle plate and being directly secured to the worktable can be more rigid, but more importantly it will be able to cope with workpiece shapes and sizes that neither of the other two methods can.

The accessories required are in most cases very simple, being T-nuts, clamp bars, studs, washers and nuts. For this purpose, the first item that comes to mind when acquiring these is to obtain one of the clamping sets (see **Photograph 4.11**). I would, though, suggest you consider carefully before going down this road; for example, if you are into small-scale model engineering, making workshop accessories, clock making, or other lightweight tasks of a similar caliber, then they are in my estimation far too large. Compare **Photographs 11.1 and 11.2**, noting how, even though they are the smallest clamps in the set, the support pieces have to hang over the edge of the table and T-slot. The answer then is to make one's own, but if

*Photograph 4.12* A pair of low profile clamps used for securing thin parts directly on the worktable. These are particularly useful where the top surface of the workpiece has to be completely machined.

you are going to use your machine largely for very heavy duty applications then they would be ideal. Even then, smaller lighter duty clamps will be preferable for many tasks, even essential when mounting items onto the angle plate.

Unfortunately, T-nuts of a given size are normally only made with a single thread size, and, as a result, the commercial T-nuts for your machine will dictate the stud size that you will have to use. Because of this, you may end up with unnecessarily large studs, nuts and washers as the photograph shows. The answer is therefore to make a set of clamping components to suit your requirements, and as will be seen later in the book their uses go beyond just being clamp bars. Some suggestions for making your own clamps, etc., are given in Chapter 8.

## LOW PROFILE CLAMPS

Sometimes when securing a part on the machine table it is not possible to use overhead clamps as the complete top surface needs machining, this being a particular problem if the part is quite thin, say 12mm (½"). In this case, low profile clamps can be employed, that in **Photograph 4.12** being a typical example. These are slipped into the T-slot and locked in position using a grub screw in the base of the slot. The grub screw is just visible behind the hexagon head. Then the screw in the hexagon head which has a cam action is used to lock the workpiece in place. They are not initially essential items, but if the need arises to machine a part in this way there are few other options so will become essential at some point. You will probably have to obtain them from a large industrial tool supplier, but see also Chapter 11 regarding possible shop-made alternatives.

# CHAPTER 5
# OTHER ESSENTIAL ACCESSORIES

Other accessories that fit into the category of being essential are very few, so this chapter is a short one.

## DIAL TEST INDICATOR (DTI)

For some set-ups the positional accuracy of the workpiece is very important, beyond that which can often be accomplished with an engineer's square and rule. This may involve accurately placing the vise or an angle plate so that the workpiece is eventually placed accurately.

To do this, the appropriate surface is tested using a DTI, as a dial indicator is unlikely to be satisfactory in many cases. Because of this a DTI should be obtained as it is far more adaptable; both are shown in **Photograph 5.1**. Because they are being used as a comparator in most cases, rather than taking actual measurements, then an analog indicator is much easier to work with than a digital one.

Mounting the indicator can present a problem as it must be stationary while the table is moving. This means that the indicator needs to be supported from a stationary part of the machine, being either the head or the column. Unfortunately, machines rarely provide a mounting point for this purpose, so it is up to the workshop owner to establish one. If the machine has a dovetail column,

**Photograph 5.2** *A facility for mounting a DTI assembly. This is mounted in place of the nut that secured the down feed stop.*

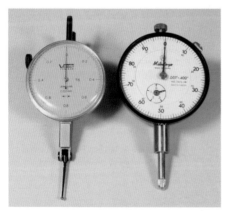

**Photograph 5.1** *A dial test indicator (DTI), left, and a dial indicator, right. A DTI is to be preferred for machine use.*

*Photograph 5.3 Accessories for mounting a DTI.*

then, having flat surfaces, a magnetic base attached to this could then carry the assembly for holding the indicator, providing that it does not have a slide protector.

In the case of a round column machine, the indicator assembly will have to be mounted on the head at some point, **Photograph 5.2**. An alternative would be to make some form of assembly that could be clamped to the round column which could then support the indicator.

Having provided a mounting point and obtained a DTI, some form of mounting assembly is essential as rarely will the

facilities provided by a magnetic base be sufficient. **Photograph 5.3** shows a typical set-up using shop-made swivel joints, but similar items are available commercially (I would suggest at least four). You can of course make the interconnecting bars, which you will need in a range of lengths.

Details for the shop-made swivel joints, together with a range of other items to complement them, are included in the book 'Model Engineers' Workshop Projects', Workshop Practice Series number 39.

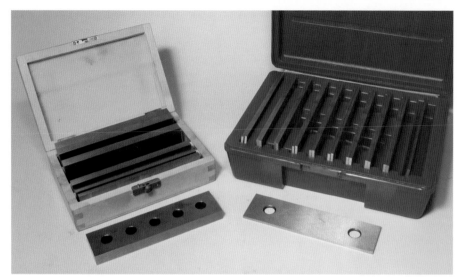

Photograph 5.4 *Two sets of parallels. That on the right is to be preferred as it has sizes in increments of ⅛" while the other is ¼".*

## MEASURING EQUIPMENT

I will assume the workshop has already had a lathe for some time and that such items as rules, engineer's squares, outside micrometers and a digital/vernier caliper already exist. This just leaves two items that, while not essential (at least not initially), are very useful.

While the caliper will have depth measurement facilities, it can be rather inconvenient in the confines of a milling machine due to its length. A depth micrometer, being compact, can often be more convenient and accurate. These come with either a fixed spindle reading just 0–25mm (0-1") or interchangeable spindles covering a wider range. Typical uses would be to measure both the width and depth of a step, or the depth of a slot being milled.

A height gauge, digital or vernier, will be useful for marking out 'machine to' lines as these reduce considerably the amount of measuring needed during the machining process. Various examples of 'machine to' lines can be seen later in the book,

typically **Photograph 11.10**. Using a surface gauge would be an alternative method, but would be far less accurate. As this item will find other uses, typically marking out components for drilling, I see this as a very beneficial item of workshop equipment, but not quite essential.

## PARALLELS

The most frequent use for parallels is in the milling vise to enable a part to be lifted so that the upper surface is available for machining. With this application in mind, their height is most critical where thin workpieces are involved. Typically, therefore, if the thickness of the workpiece is say 4mm, then the height of the parallel would need to be 2mm less than vise jaws. From this it can be seen that a wide range of sizes is required, especially so if there is more than one vise in the workshop and with differing jaw heights.

*Photograph 5.5* 321 blocks can be an useful alternative to parallels in some cases.

**Photograph 5.4** shows two sets of parallels, one having heights of 1" to 1¾" in ¼" intervals, while those in the other are ½" to 1⅝" in intervals of ⅛". The larger set is much more suited to machine work but these have the disadvantage of being only ⅛" thick which makes them much less stable. Whatever type you choose, they are an essential workshop item.

## 321 BLOCKS

These are sold in pairs (see **Photograph 5.5**), and are a form of parallel and are so named because they are 3" × 2" × 1" (metric sizes are also available). They can be very useful but are in no way essential; purchase a set when your workshop activity expands to a stage where they can be seen to be worth the expenditure.

# CHAPTER 6
# OTHER ACCESSORIES

This chapter discusses rather specialized accessories that should only be obtained when a definite requirement surfaces. Fortunately, though, when a decision is made to acquire one the range is small and the choice relatively easy (except for those with smaller milling machines as some of these accessories are quite large).

## DIVIDING HEADS
The purpose of a dividing head is to enable machining to take place at regular intervals around a workpiece. This ranges from simply machining a hexagon to take a spanner, to cutting a gear.

### SEMI UNIVERSAL DIVIDING HEAD
When obtaining a dividing head, it will be found that there is very little choice, with just the one design being commonly available and in just two sizes (see **Photograph 6.1**). These have a built-in gear ratio of 40:1 and with the provided dividing plates will achieve all divisions up to 50, but progressively less as the number increases. Even so, the available divisions should be sufficient for most situations. Unfortunately, 125, which would be required if needing to calibrate a dial for an 8TPI leadscrew, is missing.

There are minor differences in the spindle bore, taper and chuck mount dimensions, so do check with the supplier if you have a preference. This is particularly important for Myford owners as some suppliers stock them with a Myford chuck mount.

A major accessory normally supplied with a dividing head, or as an option, is a tailstock, which enables longer components to be supported. A few minor accessories

*Photograph 6.1 A semi universal dividing head.*

are also included such as a driving dog that would be used when the workpiece is held between centers.

### UNIVERSAL DIVIDING HEAD
This is a much more involved version of the above. It can achieve all divisions up to 380 by setting up a chain of gears, very like the change wheels on a lathe, to rotate the dividing plate, resulting in the greater range of divisions. Also, by driving the gear chain from the free end of the table's leadscrew, spiral flutes can be cut. All these additional facilities come at a cost with the head likely to cost as much as a medium sized mill drill.

### OTHER OPTIONS
The features that the universal head provides are most certainly beyond that required in the vast majority of home workshops. Even the semi universal head will probably be over-capable, and again it is quite expensive; also it is too large for the smaller machine. What then are the alternatives?

## SPIN INDEXER

For a simpler dividing device, the most common is the spin indexer as seen in **Photograph 6.2**. This has a fixed dividing plate with a single row of 36 holes giving divisions at 10° intervals. There are, however, 10 holes spaced at 9° to support the pin that secures the plate in the required position. By selecting the holes to use, one-degree increments can be achieved using the same principle as a vernier caliper. Even so, while giving 360 divisions spaced at 1°, it only provides 22 usable divisions, but includes most of the low numbers that may be useful (typically 6 for machining a hexagon, or 4 for a square).

*Photograph 6.2 A spin indexer that can be set at one-degree intervals but gives just 22 divisions; these include the useful low numbers 2, 3, 4, 6, 8, 10 and 12.*

Another disadvantage is that they use 5C collets to secure the workpiece. Of course, you do not have to purchase all sizes, only those that you require as the need surfaces. Also, there is no facility to mount a chuck or faceplate to secure larger workpieces; in this case, though, an adaptor based on a 5C shape would be relatively easy to make.

If the reader only anticipates simple tasks, such as milling a square or hexagon to take a spanner, or a few holes on a pitch circle diameter when securing a cylinder cover for a small steam engine, then a spin indexer would be ideal.

The indexer seen in **Photograph 6.2** has an adaptor to take also ER32 collets with the nose threaded to take the closing nut, but at the time of writing it is only available, to my knowledge, from one supplier.

*Photograph 6.3 5C (left) and ER (right) collets.*

The reader may question why there is so much projecting at the rear of the device. With the locating pin removed, this permits the spindle to be freely rotated while at the same time being able to be moved backwards and forwards. Typically, this would be useful for sharpening the helical cutting edges of an end mill; it is, however, a heavy item and would only be appropriate to be

used on an industrial caliber tool and cutter grinder. Other than that I cannot visualize any other use for it.

## COLLETS

Having mentioned 5C and ER collets, a brief explanation may not be out of place for the reader not conversant with them. **Photograph 6.3** shows one of each type with the 5C on the left and the ER collet on the right. 5C collets are very common but each collet is only capable of gripping a single diameter; as a result, one is required for each imperial and metric diameter, that is in increments of ½" and 1mm.

The ER collets are made in a range of body sizes, but for any one body size the collets are made in increments of 1mm. They have two advantages: their shallow taper makes them suitable for securing cutters and their multi slitted construction enables them to close over a range of 1mm. As a result, typically, a 7mm collet will grip any diameter between 6 and 7mm. Imperial collets are not therefore required.

## SHOP-MADE

A shop-made dividing head is an alternative approach and particularly appropriate if it is to be used on a small milling machine. In this case, a simple head such as that in **Photograph 6.4** would be worth

considering. Using the change wheels from the lathe, in increments of five teeth, this can achieve all the important low numbers and many higher ones, including 125, 200 and 360. It is shown with a gear chain, but can also be set up using just a single gear (see **Photograph 12.5**).

If you would like a more challenging project, and for a shop-made dividing head that is much more adaptable, then the one shown in **Photograph 6.5** is worth considering; by using a change wheel in place of the wormwheel used in the commercial item, it is not limited to a ratio of 40:1. Because of this, additional divisions are possible compared to the semi universal head, including the important division of 125. The worm/wormwheel assembly can also be easily removed, and with a few simple

*Photograph 6.4* A simple shop-made dividing head.

*Photograph 6.5* An involved shop-made dividing head that provides more divisions than the semi universal head as it is not limited to a 40:1 ratio.

attachments direct dividing is possible using either a dividing plate or a gear wheel. The design drawings for these shop-made heads can be found in the book 'Dividing', Workshop Practice Series number 37.

## ROTARY TABLES

While not a vast range, there are more of these available compared to dividing heads and, most importantly, they come in a wider range of sizes. This makes it easier to find a suitable table for a smaller machine. **Photograph 6.6** shows an example with a 150mm (6") table, but sizes as small as 100mm (4") are readily available. Rotary tables can in some way be compared with the dividing head in that the manual input rotates a worm which drives a wormwheel coupled to the table. There are, though, different ratios

**Photograph 6.7** *Some rotary tables can be fitted with division plates enabling them to carry out the same functions as a dividing head.*

**Photograph 6.6** *A 150mm (6") rotary table.*

available: 40, 60 and 90:1. I am unaware of any overriding factors for choosing one ratio over another and cannot therefore advise with certainty. Of course, the large 90:1 ratio would enable the workpiece to be fed more gently and could be an advantage in machining workpieces in difficult situations. Also, the calibrated handle will be marked in smaller increments.

A major feature of some rotary tables, though often supplied separately as an option, is to fit dividing plates, giving the table the features of a dividing head (see **Photograph 6.7**). While not as convenient as a dividing head, it is worth considering, having quite a cost saving over purchasing two separate items. It is worth noting that a rotary table with dividing attachment will perform dividing tasks relatively easily, but a dividing head will not be that good at rotary table tasks. Some rotary tables can be

purchased with a tailstock, enabling them to be used for long parts.

The main purpose of a rotary table is to machine curved surfaces – maybe a curve on an outer or inner surface or a curved slot. **Photograph 6.8** shows an excellent example of the end of a con rod being machined, and also shows how a smaller rotary table can be used even on the cross slide of a lathe, indicating that they can be used on the smallest of milling machines. The one shown is a shop-made item, but similar sized tables are available commercially.

## 5C COLLET FIXTURES

I include these devices (see **Photograph 6.9**) for completeness, but feel their use in the average small workshop is limited. If, however, you already have the 5C collets, the fixtures are not overly expensive and one may be worthwhile for the occasional task. When machining round components on the milling

**Photograph 6.8** A small 100mm (4") rotary table being used to machine the end of a con rod.

**Photograph 6.9** *Two fixtures that use 5C collets for holding round workpieces.*

**Photograph 6.10** *A boring head with interchangeable Morse taper shanks.*

machine, holding them securely is not always easy but providing a length is available to be held in the collet then these fixtures will perform the task with ease. The fixture on the left has a dead length mechanism which holds the collet stationary, while the internal taper moves to secure the workpiece, useful for second operation tasks, though rarely needed in the home workshop.

## BORING HEADS

These are used to produce large round holes in workpieces that are too large for fitting onto the lathe's faceplate for performing the task and are an item that in most workshops will find little use. Even so, when an application presents itself, in many cases there will be no alternative other than to use a boring head. As with the other major accessories I would advise delaying the purchase until a need for the device definitely arises.

Boring heads vary little, other than in size, and even in this respect the smallest commonly available are still oversize for the smallest milling machines. Available heads will have either two or three Morse taper or R8 shanks; in some cases they are easily removable and two shanks are provided, size 2 and 3 Morse (see **Photograph 6.10**). However, some mill drills, and maybe other machines, have a projection from the Morse taper to engage with keys on a tipped

face cutter, ensuring that the taper does not have to provide the machining force. In this case, if the taper is not long enough, the body of the boring head will prevent it from entering fully. If your milling machine is like this, do check carefully before making a purchase.

Some boring heads will be provided with cutters, which may be HSS or carbide tipped, both having their advantages. HSS is easier to sharpen and carbide is more suitable for boring castings. Sets of boring tools can be purchased separately and are not limited to being used in a boring head; with a simple adaptor they could be used as a conventional boring tool while mounted on the top slide of the lathe.

## EDGE FINDERS

Edge finders are used to accurately position the machine spindle to some aspect of a workpiece, often an edge, hence the name. There are two common types, wigglers and electronic. The wiggler is more adaptable and is my preferred version, while the electronic finder just locates an edge and when found a lamp lights. They are by no means essential but very useful occasionally. See Chapter 15, where they are discussed in depth.

# CHAPTER 7
# INSTALLATION

Fortunately, installing a milling machine is relatively simple compared to a lathe which can twist if not bolted down correctly. Of course, they are very top heavy machines and effort should be made to reduce the center of gravity by lowering the head and knee prior to moving. Work safely and keep in mind that once the machine tips, there is no stopping it. Never put yourself into a position in which you are unable to get out of the way should the machine tip.

Remove all items that can be removed easily; this is only likely to achieve a limited reduction in weight, but if you wish to dismantle further then consider the options very carefully. Endeavour to find some help with the task and if there is room in the workshop hire an engine hoist or similar.

If bench mounted, and I feel sure most will be, the bench must be strong enough to take the weight. In addition to ensuring that it can take the weight, do also brace the structure to ensure that it is rigid and unable to swing side to side or back to front. If this requirement is not met, then the machine may vibrate when in use, and if nothing else this may affect the quality of finish. Anchoring the bench to the workshop's structure is also worth considering.

While by no means essential, do position the machine so that its table surface is level, at least to the precision of an average spirit level; acquiring a precision level for the purpose is unnecessary. The reason for this is that workpieces that need machining at an angle (not precise) can be set using the protractor head of a combination square set.

Do, when siting the machine, ensure that there is adequate room for the table to move

**Photograph 7.1** *A mill drill set into the corner of a workshop can sometimes make best use of the available space.*

fully, both left and right. This may seem obvious, but if bench space is limited then setting the machine at an angle at its end may make better use of the space available (see **Photograph 7.1**).

If you are installing the machine in a workshop with minimal headroom, it may be difficult to remove the draw bar, which is sometimes necessary, as some accessories have different threads in the end of their shank. Even if this is not the case with the initial accessories acquired, it may become

necessary when an additional item is obtained later. Of course, one can remove the items from the worktable and the cutter chuck so that lowering the head will give maximum headroom. If a problem may exist, then installing the machine under the highest point of the roof may be the only option, even if less convenient when other factors are considered.

Electrical installation is simple, as the normal electric socket will be able to supply more than enough power; do, however, install the socket local to the machine to avoid a long cable run, and even then clip the cable as far as is possible in a secure location so that it is unlikely to be damaged. If you are installing an old machine, it is unlikely to be fitted with a no volt release switch, so do obtain one and include it in the installation. A direct on line starter with overload protection would be even better.

*Photograph 7.2 Some milling machines have rotating heads. On installation, these should be checked to see that the spindle has accurately been set upright when supplied.*

*Photograph 7.3 Checking to see that the machine spindle is upright relative to the machine's worktable. The error should be no more than 0.025mm over 300mm (.001" over 12").*

*Photograph 7.4 Making a small adjustment to the head's rotation. See text for an explanation.*

## ROTATING HEADS

Some machines come with heads that can rotate, enabling the cutter to approach the worktable at an angle (see **Photograph 7.2**). In this case, do check that it has been accurately set upright using the set-up shown in **Photograph 7.3**. With this, the machine spindle is rotated by hand and the readings noted at both ends of the table. The international standards for machine tools state that the difference should be no more than 0.025mm over a distance of 300mm (.001" over 12"). Incidentally, this applies to fixed head machines but obviously is a value worth aiming for on machines with rotating heads.

If a greater error is found, making very small adjustments can be a tedious process of trial and error, but this can be largely eliminated by the set-up in **Photograph 7.4** which has a small plate with a drilled center in it fixed to the table. A center is then fitted into the machine's spindle and lowered into the drilled center in the plate when very small adjustments can easily be made by traversing the table left and right. See Chapter 16 regarding the effect of the spindle not being perpendicular to the movement of the machine table.

While not actually a part of the machine's installation, do ensure that a method of having all the required tooling is available locally to it. This will include such items as spanners for the clamping accessories, drill chuck key, and so on. A partial example of this can be seen in **Photograph 7.1**, right side and below.

# CHAPTER 8
# GETTING STARTED

With the machine, and the chosen accessories, acquired and physically installed, you are now ready to start the process of actually machining. The obvious choice, if new to the machine, is something simple. If you followed my suggestion in Chapter 4 and have decided to make your own table clamping components, then these are an ideal first project. However, while I am including here the set-ups, etc., necessary to make these initial items, do read on to Chapter 16 before actually starting machining.

If the workshop already has the required clamping equipment, then the reader can bypass this chapter.

## T-NUTS

If you have access to the standard dimensions for T-slots and T-nuts for your machine, then work to these, except use your chosen thread size, whether metric or imperial. If you do not have these details, then the following will be more than adequate. Make the main body of the nut a little under the width of the slot in the table, say 0.5mm (.020"), and the height 2–3mm (³⁄₃₂"–⅛") less than the space in the table. Make the width and thickness of the T 2mm less than the table dimensions, the space enabling the nut to slide easily in the slot even when small fragments of swarf are present. The length of the nut can be made 5mm greater than the width across the T. It is probable then that, having chosen these sizes, they will not conform to a standard material size, but providing the difference is not great then amend your chosen sizes to suit the material available – the width, height and length are not critical.

Take a length of material sufficient for at least six nuts and mark out and drill tapping size holes for the chosen thread. As a guide, I have made nuts with both M8 and M10 threads, but the M10 T-nuts rarely get used unless I run out of M8 T-nuts. If your vise is one of the smaller ones, you will have to start with two pieces of steel, each making three. However, before you can machine these, two temporary T-nuts will have to be made manually (saw and file task) so that you can secure your vise to the machine table.

Place the material into the vise using parallels to pack it up, and machine one side of the T-nut central leg (see **Photograph 8.1**), followed by turning the workpiece round and machining the second side similarly. If you have had to start with oversize material, turn the workpiece over, holding it on the

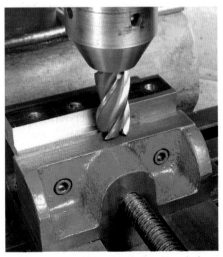

*Photograph 8.1 Machining a length of steel to make four T-nuts.*

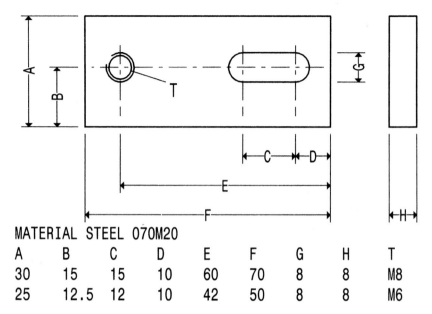

MATERIAL STEEL 070M20

| A | B | C | D | E | F | G | H | T |
|---|---|---|---|---|---|---|---|---|
| 30 | 15 | 15 | 10 | 60 | 70 | 8 | 8 | M8 |
| 25 | 12.5 | 12 | 10 | 42 | 50 | 8 | 8 | M6 |

*Sketch 8.1 Suggested sizes for workpiece clamps. G is quoted as 8mm, but should be changed if a large stud size has been chosen. Make at least four of each size.*

central leg, and machine the base to establish the required dimension. The ends of the T can also be machined at this stage if required, but you will need to lift the workpiece a little so that you can carry out this operation without contacting the top surface of the vise. Tap the holes already drilled and cut into individual nuts, followed by machining their ends while held in the machine vise, or on the lathe in the four jaw chuck.

## CLAMP BARS

Next task is to make some clamps (I would say at least four) to the size you have chosen. I suggest making two sizes, though, as you will need smaller ones when working with the angle plate and may also find a use on the table when smaller workpieces are involved (see **Sketch 8.1** for suggested dimensions). Cut sufficient lengths and place the pieces in the vise as a pack and machine their sawn ends (see **Photograph 8.2**). Next, one at a time, machine the slot using the table's stops

*Photograph 8.2 Machining, as a group, the sawn ends of some workpiece clamps.*

to set the length. If you align the end of the clamp with the end of the vise jaw each time, then the end stops will not need resetting for each clamp (see **Photograph 8.3**). Drill and tap the hole, the purpose of which will become evident later in the book.

## STUDS

Next you will need some studs, but rather more than the T-nuts as you will need them in differing lengths; I would suggest 40, 60, 80, 100 and 120mm (1½", 2⅜", 3", 4", 4¾"), making at least two of each. Commercial studs will sometimes be threaded at one end

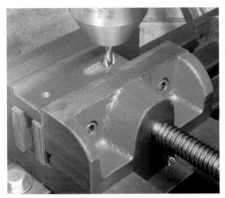

**Photograph 8.3** *Milling the slot in a single workpiece clamp.*

just to the depth of the T-nuts, so that the stud does not pass through and contact the base of the T-slot. The stud is then threaded from the other end to take the clamping nut. If made like this it would be a very large task, and I would suggest acquiring a length of studding and cutting it to the required lengths. To avoid the stud passing through the nuts you have made, take a pin punch and distort the thread at a few points around the base of the thread.

## NUTS AND WASHERS

While standard nuts will suffice, you will need to make some large washers for use when clamping up to devices with slots, typically the angle plate. For these I would suggest 20mm × 3mm thick for M8 studs and 25mm × 3mm for M10 studs, or 2½ times the diameter of your imperial threads and ⅛" thick.

**Photograph 8.4** shows a typical set of shop-made clamping accessories which, having an M8 thread size, are perfectly adequate for the

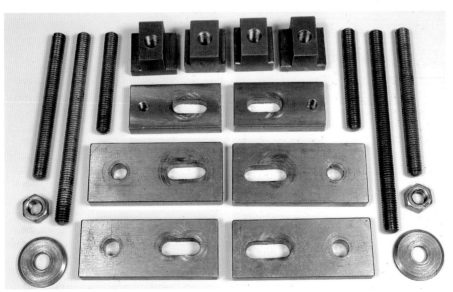

**Photograph 8.4** *A set of shop-made workpiece clamps.*

**Photograph 8.6** *Clamps can often be used as fences or fixed jaws for use with low profile clamps.*

**Photograph 8.5** *Having made the clamps, keep them easily available at the side of the milling machine.*

majority of tasks that will be performed in the average small workshop. Having made your clamping accessories, do give them a good home adjacent to the milling machine (see **Photograph 8.5**).

With experience in their use, you will probably find other sizes useful. I use some that are rather wider and shorter in length (see the top row of the photograph). Frequently, these wider clamps are used as supports against the edge of a workpiece rather than actually clamping it, or as fixed jaws for low profile clamps (see **Photograph 8.6**).

This brings me to one final point for this chapter. With the wide range of workpiece shapes and sizes, you will often find it beneficial to make special clamps, supports or fences for the task in hand, especially when using the angle plate. Do not attempt to bypass this, as having made the items initially you will find many other uses for them in future projects.

# CHAPTER 9
# USING THE VISE

When a task for the milling machine surfaces, it is logical first to consider a vise to hold the part. However, this I consider results in the vise often being used when other methods would be preferable, a factor that I will seek to expand on through the coming chapters.

The main advantage of the vise is the speed at which the workpiece can be clamped and released, a factor that would appear very beneficial where a batch of identical parts are to be machined. Unfortunately, this is frequently not the case as batch production invariably requires subsequent parts to be placed in exactly the same position. Parallels below the part would satisfy one axis, with the fixed jaw another. The third axis, along the jaw, is, however, more of a problem.

To overcome this, end stops are available, but there are only a limited number of types available so finding one to fit your vise may be a problem. While these can be useful, they would not provide for parts that had to be placed in the vise at an angle or where a narrow part had to be set perfectly upright. Having emphasized batch production, I include in this the small quantities that often appear in the home workshop; even positioning one-offs can often present a problem.

Another factor when considering using the vise is that the depth of its jaws are relatively shallow and do not cope well with thin parts that would have to project above the jaws appreciably. Even so, if a decision is made to use the vise it will now need to be positioned on the worktable.

## POSITION ON THE TABLE
Some milling machine operators consider that the force involved in machining a workpiece should always be towards the vise's fixed jaw, and I can see some reasoning for this, especially when being used on the horizontal milling machine. However, in the case of vertical milling it is frequently necessary to complete a part without removing it from the vise and requiring machining to take place in different directions, perhaps occasionally in all four – left-right, towards-away. From this, it is obvious that it is far from an essential requirement.

I normally place my vise along the length of the table with the vise's handle on the right, which, being right-handed, is logical for tightening the vise onto the workpiece. There are, though, occasions when, typically, a long part requires machining centrally but one end will collide with the mill's column. In this case, the vise has to be mounted with its jaws parallel with the table's traverse.

## PLACING THE VISE
The process of positioning the vise on the table will depend on the accuracy required for the task about to be undertaken; for example, if the part needs to be reduced in thickness over just the top surface then accuracy is of no importance, so visually in line will be more than adequate.

For some parts, however, precision will be the requirement and the method shown in **Photograph 9.1** will have to be adopted. In this the DTI is placed against the inside of the vise's fixed jaw and the table traversed, adjustments to the vise being made until the indicator reads the same across the complete

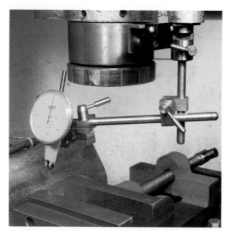

**Photograph 9.1** *Setting a vise accurately in line with the table's traverse.*

**Photograph 9.2** *Using a piece of bar in the vise with two 321 blocks enables a vise to be accurately placed using an engineer's square.*

length. The disadvantage of this is that it is initially time-consuming to set up and then time-consuming to place the vise, but it must be done if the part being made demands it. Some will suggest that the method should be used every time a vise is placed, because a subsequent project may require this level of accuracy and if it has been placed to a lesser accuracy then this may be forgotten at the time. The reader will have to decide whether this is appropriate to his or her situation.

I rarely require that level of precision, and the vise will have been fitted and removed very many times before a need arises. I therefore condition myself to reset the vise, when a high level of accuracy is finally called for; of course, the vise may not even be on the table at that time. This approach does depend on a quick and adequate method to bridge the gap, between visually and precision.

One method is to place a piece of rectangular bar in the vise with its ends projecting, and with a 321 block at either side use an engineer's square to set the vise's position (see **Photograph 9.2**). An easier and

**Photograph 9.3** *If the vise is suitable, then an engineer's square directly off the end of the vise is a very quick and accurate method of placing a vise on the worktable.*

potentially more accurate method is to use the engineer's square directly against some part of the vise (see **Photograph 9.3**). If you have purchased a toolmaker's vise, or something similar, then the front face should already be parallel with the vise's jaws. If this is not the case, for whatever type of vise you have, then you have to machine a strip across the front of the vise using the following method.

Clamp a bar onto the milling machine table with it running front to rear, and set this precisely using the method shown in **Photograph 9.1**. With that done, turn the vise upside down and secure it onto the bar; the front end can now be machined so that it is parallel to the jaws. It is not necessary to machine the end of the vise completely, only a narrow strip at its base, say 5mm (.2") wide. Having done this, mount the vise on the table using a square to position it and check the result to confirm that the process has been a success. Subsequently, mounting the vise will be a rapid operation and accurate enough for all but the most demanding applications and will repay many times over the time taken to machine the vise. With the method being so quick and easy, positioning the vise visually will not be necessary even for the least demanding applications.

### USING PARALLELS
A frequent feature of using the vise is for the workpiece to be too thin to sit on the bed of the vise while still allowing the top face to be machined. To overcome this, the workpiece will have to be raised above the bed assisted by a pair of parallels (see **Photograph 9.4**). The parallels in some sets, typically those

*Photograph 9.4 Parallels are required when machining the top of a very thin workpiece.*

*Photograph 9.5 Using a piece of thick rubber to keep thin parallels upright and against the vise jaws.*

seen in **Photograph 5.4**, right, are only 3mm (1/8") wide, resulting in them tending to fall over when the workpiece is placed in position. To overcome this, place a strip of rubber between the two parallels to keep them against the vise jaws (see **Photograph 9.5**).

### JAW LIFT/TWIST
Jaw lift was covered in Chapter 4, but unless the reader has obtained one of the very heavy duty vises a very small amount is inevitable. In this case it may be able to minimize the effect of the lift by tapping the workpiece near the moving jaw with a soft hammer after the part has been secured. Lift only occurs, however, if the majority of the workpiece is above the axis of the clamping screw.

Jaw twist is less critical and can in fact be advantageous if the faces of the part being machined are a little off parallel. However, even if less critical it is essential that the user is aware of the problems that can result and the precautions that should be taken. **Sketch 9.1A** shows a part having been placed in the vise so that the end face of the part can be surfaced using the side of an end mill.

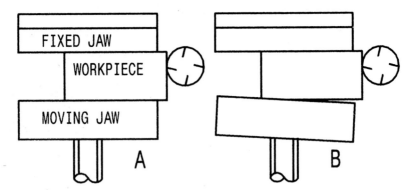

*Sketch 9.1 Effect of jaw twist: (A) lightly closed; (B) firmly closed.*

Even though a small part of the workpiece extends beyond the clamping screw's axis, this end will see the greatest pressure and will compress the most causing the jaw to twist when tightened, as **Sketch 9.1B** shows, but much exaggerated. Because of this, the clamping force will fall away considerably towards the jaw's edge, resulting in a possibility that the part will lift due to the helical nature of the end mill's cutting action.

This problem is not confined to economy vises and may still occur with more expensive ones. To avoid the problem the part should be held so that the clamping force is not far off its center, certainly no greater than a $^{60}/_{40}$ difference. This would of course eliminate the possibility of machining the end face as above.

Sometimes a shorter part will need to be held at the end of the vise jaws for some reason and a similar problem will result. In this case, a similar thickness packer placed at the other end will ensure that both parts are securely held. A packer may have helped in the example in **Sketch 9.1**, but only if it was exactly the same width or preferably very slightly thicker. A piece of the same bar material plus a piece of paper would be ideal.

## POSITIONING THE WORKPIECE

It has already been mentioned that a major disadvantage of using a vise is that it is difficult to position the part precisely and some other method has to be used, normally an angle plate. If, however, all that is needed is for the part to be set upright in the vise, then an engineer's square can be used. In some cases the part may be either too thick so that the square falls between the jaws, or too wide so that there is insufficient jaw available to support the square. As an alternative, a cylindrical square will perform the task easily (see **Photograph 9.6**). Unfortunately, these are only made in sizes that would be too large and are far too

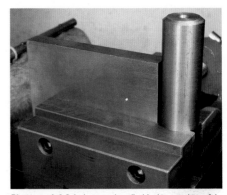

*Photograph 9.6 A shop-made cylindrical square is useful for placing a part upright in the vise.*

costly as they are an inspection item, but making one sufficiently accurate for the task is very easy.

Place a length of steel in the three jaw, say 30mm diameter by 80mm (1 ⅛" x 3") long, and skim the diameter over the available length; use a large center drill to relieve the center and face the end, then part off. While the diameter is unimportant the result must be parallel; if it is not, then this may be the time to adjust the lathe's mountings. With the cylinder completed the sides will be square with its faced end. Having parted off, face this end and add a substantial chamfer so that it is known which is the working end. You can even use it to check your smaller engineer's square.

A similar task is to place a piece of metal in the vise but at an angle other than at 90°. In this case, either a combination square or a universal bevel protractor would seem appropriate but positioning the device onto the vise and its arm against the workpiece is unlikely to be possible due to the bulky nature of these devices. One method of overcoming this is to use just the protractor head of the combination square and set it to the required angle, then use the spirit level in this to indicate when the angle is as required (see **Photograph 9.7**). An alternative is to use one of the modern digital protractors, some of which have magnetic bases so that it can be attached to the part being positioned. Of course, neither method can be considered to be a precision method, but should be adequate for many applications.

While a relatively easy method to use, there will be an error if the machine table is not level. To overcome this, place the protractor on the machine table and set it to read level. From this read off the error and add this to, or subtract it from, the angle required and then set the protractor to the resulting value. Using

*Photograph 9.7 The protractor head from a combination square, while not precision, will often be adequate when requiring to place a part at an angle in the vise.*

*Photograph 9.8 When holding castings in a vise, pieces of soft copper or aluminum placed between each jaw and the workpiece will improve its security appreciably.*

the digital version will be easier as it can be placed on the machine table and zeroed. However, carrying out the task using an angle plate will be far easier and more accurate and definitely quicker if more than one part is being made, and it will also better support a longer part.

## HOLDING IRREGULAR ITEMS

When faced with holding an irregular item, frequently a casting, a problem will result due to the faces requiring to be held not being parallel. While in some cases the tapered nature of the workpiece will be too great for a vise to cope with, the taper present between the faces of a casting may be small enough in others to make it a possibility. Even it these cases, just placing the casting directly into the vise is not advisable and it should be held between pieces of copper or aluminum. These, being soft, will conform to the surface of the casting, thereby improving the grip considerably (see **Photograph 9.8**).

## HOLDING ROUND ITEMS

Using the vise to hold round items is in some cases perfectly acceptable, and in others needs careful consideration before attempting it. Typically, longer and/or horizontal parts will be more at home in the vise than will those that are shorter and/or vertical. In all cases I would suggest that the length of the part should at least be equal to its diameter in view of there only being a line contact between part and vise.

Placing a round item vertically in the vise unaided is best avoided, but this is easily overcome by using a small V-block as shown in **Photograph 9.9**. While this shows just a thin part supported by a parallel, it would be equally acceptable with a longer length. One word of warning: it is possible to split the V-block in two if you are over-enthusiastic when tightening the vise.

*Photograph 9.9* Using a small V-block enables a round part to be securely held in a vise.

*Photograph 9.10* A budget tilting vise is being used to hold a plate so that a rectangular hole can have an angular face.

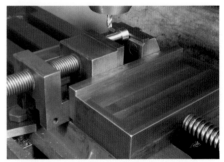

*Photograph 9.11* When a very small part has to be held, using a vise-in-vise method is worth considering, as the part can be fitted away from the machine. This is particularly useful if the part needs to be held at an angle.

## USING A TILTING VISE

These follow much the same rules as for a fixed angle vise. Do remember, however, that you need to position the vise in line with the table's axis before adding the workpiece and setting its angle. **Photograph 9.10** shows a typical application.

## VISE IN VISE

Positioning a very small part in a large vise can present a problem and the vise-in-vise set-up is worth considering if you have two suitable vises (see **Photograph 9.11**). The main advantage is that the part can be positioned in the vise away from the machine, often easier with parts so small. It is particularly helpful when needing to position a small part at an angle as the part can be held normally in the small vise and then the vise set at the required angle, a method worth considering even for larger workpieces.

# CHAPTER 10

# USING THE ANGLE PLATE

Having stated in the last chapter that the advantage of the vise was its ability to rapidly secure and release the part being machined, the advantage of the angle plate is that it is capable of working with a much greater range of workpiece shapes and sizes. Also, it is frequently easier to precisely position the workpieces, many of which would present a difficult or even an impossible task for the vise. Even though an application for the angle plate may take longer to set up it will, once set up, frequently rival the vise for speed if a batch of parts is to be made. Also, repeatability is much more easily achieved using an angle plate, often being near impossible if using a vise. Another important factor is that the much greater depth of an angle plate, compared to a vise jaw, enables them to support larger and less robust parts.

## POSITIONING THE ANGLE PLATE
This, like the vise, depends on the level of accuracy required, but for the majority of tasks a square off the edge of the machine table and its blade against the angle plate's face will be adequate. If a greater level of precision is required, then check its face using a dial test indicator, in the same manner as would be done when placing a vise.

Unfortunately, with the angle plate being able to support a very wide range of workpiece shapes, I can only provide a few examples of its use.

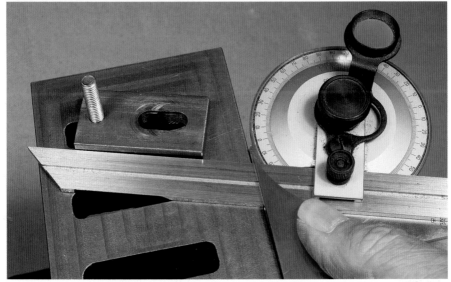

**Photograph 10.1** *Positioning a fence on an angle plate to be used to set the angle of the workpiece to be machined. The angle plate is being held horizontally in the bench vise, which is much easier than carrying out the task vertically on the machine table.*

## EXAMPLE 1

If a part is required to have an angled end and with a notch added, it could easily be set in the vise using the method suggested in the last chapter, but the machining would have to take place much too far above the vise jaws. It is therefore an application for the angle plate. While not visible in the photographs, the angle plate is being held in the bench vise with the working face horizontal, always a good idea as it makes it much easier to position items, similarly a good idea when using the lathe's faceplate.

For this task, therefore, a fence was fixed onto the angle plate's face to position the part accurately at the required angle (see **Photograph 10.1**), and with that in position a clamp was added to secure the part to be machined (see **Photograph 10.2**). Actually a second clamp would have been worth adding for additional security.

With that done, the assembly was transferred to the milling machine and the angle and notch made (see **Photograph 10.3**). If more than one part were to be made, then adding an end stop would also have been a good idea, as subsequent parts could then be added and removed with ease and without the need to position the workpiece or to measure the work being undertaken. While I emphasize the benefit in the case of batch production, I realize that the home workshop will rarely make largish quantities. Even so, quantities of two, three or four are not uncommon, and even at these quantities the approach is very worthwhile. Due to the length of the part being machined, this would have been very difficult to carry out with the part in the vise, though some may have sandwiched it between two thicker pieces of steel. Setting the angle would still be less than easy.

*Photograph 10.2 The workpiece is added.*

*Photograph 10.3 The assembly is transferred to the milling machine and the part machined.*

## EXAMPLE 2

Again I am using two fences, but the reasons are different. In this case the lower fence, just seen in the photograph, sets the part horizontal so that the recess being machined would be parallel with, and the same distance from, the part's base as a number of parts were being made (see **Photograph 10.4**). However, as the recess passes right through the part, the fence on the left would seem superfluous even for repeat parts – there is, though, an important reason for its inclusion.

Photograph 10.4 *A typical set-up using a single clamp and a supporting fence.*

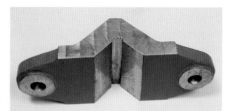

Photographs 10.5 *(above)* **and** *10.6 The part in* **10.5** *is clamped to a square block on the angle plate set at 45° (10.6).*

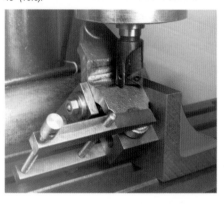

A major requirement when clamping parts to the angle plate or the machine table is that sufficient clamps should be employed, and, to repeat a comment that I often make, 'one too many is better than one too few'. In some cases, though, this cannot be achieved and fences that just support the part are often used; the fence on the left is there for this purpose.

I realize that this is a task that could easily be done using the vise, but I include it as it illustrates the use of fences to support as well as position. In any case, there were four parts being made, each needing machining similarly and on opposite edges, so the time taken to set up was minimal compared to the machining time taken. Also, there were no loose parallels to remove, so they and the vise base could be cleared of swarf between parts, the fixed fences being much easier to keep clear, something that I often find to be a major benefit when using the angle plate.

## EXAMPLE 3

This example shows a rather more complex part, being a casting that requires a V machined into it, **Photograph 10.5** showing the finished part. Even though precision was not called for, making the V reasonably

accurate in terms of angle, depth and being central was worth attempting, if only for the experience gained. The machining in progress is shown by **Photograph 10.6** and from this it can be seen that a square block has been secured to the angle plate and set at an angle of 45°.

The lugs on the hidden side of the casting had already been surfaced and one was placed onto the block and the casting secured using a toolmaker's clamp. Obviously, this would not be sufficient and a bar clamp was added also. Note also that the cutting force is towards the angle plate. As there was no provision for the casting to return to exactly the same position for the second side measurement was again needed; it did, though, ensure that the angle was easily set up and correct for both faces.

Having shown an example using a toolmaker's clamp, a note about their use would not be out of place. When using these on the milling machine, it is essential that they are set correctly, that is that it should be impossible to swing the clamp about the point at which it is clamping the component. For example, if not correct and the clamp swings about its tip, then the two arms are too far apart; loosen the outer clamp screw and close the central one a little and test again, repeating until the clamp cannot be moved. If the clamp pivots about the inner edge of the item being clamped, the two halves are too close together; loosen the outer screw and unwind the inner one slightly, then close again using the outer screw. Repeat until the clamp is totally secure.

This process is vital when using these clamps on the milling machine as vibration due to the machining taking place may cause the clamp to swing and it could then fall into the path of the cutter – even worse, it could fall off completely, with disastrous results.

## EXAMPLE 4

Using the angle plate is not confined to larger parts, though smaller items will be in the minority. An example are the special T-nuts shown in **Photograph 10.7**. Having been made from a round bar, and the recesses already made to create the T-nut form (see **Photograph 9.9**), flats were required at right angles to the already machined recesses.

To hold these for this operation, a fence was added to the angle plate ensuring that its end was at 90° to the worktable. A second fence was then fitted so that the space was just sufficient for the T-nut to sit between them, this being achieved by holding one of the nuts between the fences as the second fence was secured.

The nut was then secured, using a screw through the slot in the angle plate and a 321 block to set its height above the machine table, and the top machined, nuts two and three being machined similarly. Next the nuts were fitted, flat side down, on the 321 block, the cutter was lowered and the second flat

**Photograph 10.7** *Two fences and a 321 block are being used to position the part (see text).*

**Photograph 10.8** *Using an angle plate mounted on its end can sometimes be a good idea.*

## EXAMPLE 5

**Photograph 10.8** is included just to point out to the reader not to forget that an angle plate can in some cases be mounted on its side on the worktable. In this case, a clamp bar is being used in one of the plate's slots (bottom left) and a bar across the two flanges of the plate (top right) to secure the plate onto the worktable.

## EXAMPLE 6

I have included this example of a small toolmaker's vise having its securing slots machined (see **Photograph 10.9**), as it illustrates the use of a parallel off the machine table. Even though it is a single part, the set-up was used four times, for the two sides and the two slots, and also, as I mentioned above, the effort involved in keeping the assembly clear was minimal compared to using a vise.

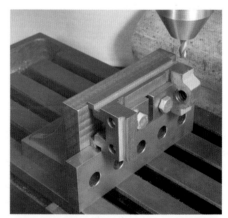

**Photograph 10.9** *Parallels can also be used with the angle plate, in this case supporting a small vise casting.*

made. The set-up made it a quick and easy task, and this is a case where a small batch of parts surfaces in the home workshop.

If held in the vise using the two main faces, it would have been difficult to ensure that the sides of the already machined T-nut were upright. To overcome this problem, they could have been held at the end of the vise jaws (suitably packed at the other end), but I was concerned about security if done this way with there being so little to grip. It should never be forgotten that security of the workpiece is an essential requirement when using the milling machine. Incidentally, I am using the term 'fence', but in most cases these are clamp bars being used for that purpose.

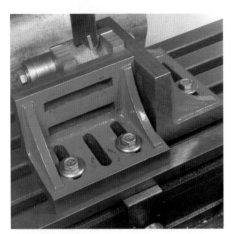

*Photograph 10.10 Using two angle plates to support a part (see text).*

*Photograph 10.11 Protecting an angle plate with a piece of hard card is a good idea when securing an unmachined casting.*

## EXAMPLE 7

In Chapter 4 I suggested that purchasing two identical angle plates could be beneficial for some set-ups. I accept that the subject in this example is an extreme one, but I could see no other easy alternative that would give me the required level of precision. The set-up uses two precision angle plates, and a precision cylindrical square (see **Photograph 10.10**).

The requirement was to machine the side of a V-block so that it would be parallel with the V already made and at 90° to the base also already machined. The first angle plate was secured to the table and the cylindrical square to that. With the V placed against the cylindrical square the side would then be parallel to the V when machined. I did also need the base of the block to be at 90° to the table. It occurred to me that if I placed my second angle plate against the base of the V-block this would ensure that it was perfectly upright, but how was I to clamp the parts together? With some thought, I decided that this was not needed and I machined the side, ensuring that the machining force was towards the first angle plate, which was done with ease. The cut was only a light one as the block had already had all the faces rough machined.

This is not a set-up that will find much use, but it does indicate how adaptable the angle plate is and that two may occasionally be beneficial.

## EXAMPLE 8

In this example, I show the first face of a casting being machined, and because of this the face against the angle plate is still in its rough condition. To protect the angle plate and improve the hold, a piece of thin hard card has been placed between the two. This can be seen at the lower right corner of the angle plate in **Photograph 10.11**.

**Photograph 10.12** *Using two cylindrical squares ensures that the end of the angle plate being machined is square to both faces.*

**Photograph 10.14** *Using two long studs as an alternative to a split vise.*

## AN ALTERNATIVE TO THE ANGLE PLATE

A pair of cylindrical squares will often make a good alternative to an angle plate, as shown by **Photograph 10.12**. In this, a small angle plate is being machined on its end while supported by two cylindrical squares, ensuring as a result that the end is perfectly square to both of its main faces.

**Photograph 10.13** shows one of the faces of the angle plate also being machined. I include this as it illustrates that if a large part requires machining and is too big for the available angle plate to hold securely the squares can be set much further apart. The squares were originally made to be used on a smaller milling machine, having smaller T-slots, and would benefit from being a little larger in diameter for uses on the machine table shown. Even so, I have used them many times without any problem.

Should the reader be interested in making one or two angle plates without the use of another, then the method was described in 'Milling a Complete Course', Workshop Practice Series number 35.

Having completed the details for the book a much more likely use for two angle plates surfaced in my workshop compared with that shown in **Photograph 10.10**. As a result I am adding **Photograph 10.14** showing that with two long studs they can be used as an alternative to a split vise. In this it can be seen that a large casting is being machined fully over its upper face, the jacking screws seen were just for supporting the part while it was being secured.

*Photograph 10.13* Two cylindrical squares being used in place of an angle plate.

## FINALLY

With there being an almost infinite number of possible set-ups, this chapter has only been able to illustrate a few. From these, however, the reader should have gained the basics for using an angle plate. For two more, see **Photographs 11.8 and 17.3**.

# CHAPTER 11
# USING THE WORKTABLE

The main advantage of using the worktable is that very much larger parts can be secured and machined. Even so, using the method is not confined to those that are too big for the available vise or angle plate. Before giving practical examples, I will cover the basics of workpiece clamping, most of which also applies to mounting parts on the angle plate.

**(1)** The packing under the outer end of a bar clamp should be just higher than the height of the part being clamped. While not critical, the amount should increase with longer clamps – typically, I would suggest, +0.5 to +1mm higher for a 50mm long clamp (.02 – .04" for 2") and +1 to +2mm for a 100mm long clamp (.04 – .08" for 4").

**(2)** Do attempt to get the clamping screw or stud as near to the workpiece as is possible, and at least nearer to the workpiece than to the packing.

**(3)** Always attempt to use two clamps minimum – three, if possible, if the task is a heavy duty one.

**(4)** Where you are limited in the number of clamps that you can fit, add supports around the base of the workpiece to increase security.

**(5)** When clamping angle plates and vises to the worktable, using screws can often be more convenient than studs. Before fitting the screw, stand it in the T-slot alongside the item to be clamped to ensure that it will not contact the bottom of the T-slot when tightened. If this were to happen, it might

be thought that the vise or angle plate is fully clamped when in fact the screw has tightened itself against the base of the T-slot. In this case it may move while machining is taking place.

**(6)** Always use very heavy duty washers (at least ⅛" thick) when clamping parts with slots rather than round holes.

## PACKING TYPES
The three main types of packing for the outer end of the clamp bar are: stepped blocks, pieces of stock material, and jacking screws.

Stepped blocks are provided with the clamping sets seen earlier (**Photograph 4.11**), although they can be purchased individually. The method is used in two ways: one where there are two blocks and the clamps sits on these, and another that uses a single stepped block and the end of the clamp mates directly with that (**Sketch 11.1** illustrates the principle). **Photograph**

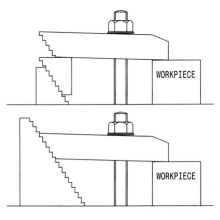

*Sketch 11.1 Using stepped blocks and clamps.*

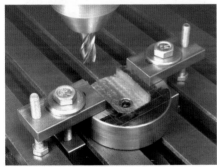

*Photographs 11.1 (top) and 11.2 Commercial and shop-made workpiece clamps compared.*

*Photograph 11.3 Using a piece of packing equal to the workpiece thickness works well, but the height should be increased by a thin piece of card or plastic to provide the required increase in height.*

**11.1** shows a set-up using these which should be compared with **Photograph 11.2** which was the actual method I used for machining the part. They are the shortest clamps in the set yet still hang over the end of the worktable at one end and the T-slot at the other.

Stock material is an effective method if used individually, but do not attempt to use a pile of random sized pieces of steel to create the packing required. However, while I said a single piece, this will often be the same thickness as the workpiece, so adding a thin piece of hard card or plastic will provide the required difference in height (see **Photograph 11.3**). When making more than one part, one can often be used as the packing while the other is being machined.

Jacking screws, being continuously variable, overcome the problem of finding the correct packing, especially if the part to be clamped is some random size and using a piece of stock material cannot be adopted. The essential requirement is for the clamp to have a hole to take the screw at the outer end. While the principle is the same, there are a number of alternatives which also depend on whether the hole is a clearance or tapped. If the hole is a clearance one, then the jacking screw will need securing with a nut on either side; with a tapped hole, the screw can be just threaded into the clamp, although a nut locking the screw on one side will provide a little more stability.

Rather than using a screw, a stud can be used with the lower end being threaded into a T-nut which can then either be placed in a T-slot or stood on the table's surface. I am not including any examples here as the method is used in a number of the photographs that follow. Having covered the basics, I will now provide a few examples.

**Photograph 11.4** *A useful set-up when needing to machine a quantity of identical parts to the same length.*

## EXAMPLES

The examples given in this chapter are only a very small number of the possible methods to use when clamping workpieces to the machine table, but should give the reader an adequate insight into the method. Do remember: security of the workpiece is vital!

## EXAMPLE 1

This is a very simple example showing the end of a short piece of rectangular bar being machined to length (see **Photograph 11.4**). If this was just a single part and the vise was already on the machine table, then using the vise would be the obvious method. In this case, though, a number had to be made to the same length. As not totally obvious from the photograph, see **Sketch 11.2** for the details of the assembly; even so, further clarification is needed.

The assembly appears not to be following my advice in (3) above, but does have the fence which will support the part. See also the end stop (not shown on the sketch) that sets the length and provides a limited amount of additional support. Most important,

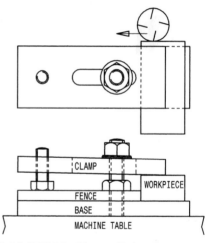

**Sketch 11.2** *Details of the assembly shown in* **Photograph 11.4**.

though, is that the workpiece projects beyond the fence by no more than a millimetre, minimizing the tendency for the part to be levered out from under the clamp. Once set up, it was much quicker than using a vise, as subsequent parts could be made without the need for measurement.

## EXAMPLE 2

This shows a round item, 60mm (or 2⅜") diameter, secured to the machine table and a recess being milled into it (see **Photograph 11.2**). Some may have attempted to carry out this operation with the part held in a vise, though holding round items in the vise is far from a good idea at this diameter, particularly taking into account the load that the machining would place onto it and the limited depth available to be placed in the vise. Incidentally, the part then had dovetails machined on either side of the recess while still on the table.

The jacking screws made the set-up easy as there was no searching for packing and it was very secure; even so, the part is also being held with a central cap screw into a T-nut below, although I am sure this was unnecessary. If you set yourself up to use this method of clamping, even if only occasionally, then do reserve screws for the purpose and machine away the raised lettering on the head as this may make marks in the worktable. Note also the heavy duty washers being used under the screw heads.

## EXAMPLE 3

In this example, **Photograph 11.5**, a much larger round item is being secured and is located onto a round post fixed to the table. This enables the part to be rotated for each slot milled. The photograph shows the final stage where a slot is being made into a T-slot. Note that the stud is nutted into the T-nut and passes through a clearance hole in the clamp bar with nuts on either side. The nut below the bar, unseen, sets the height and one on top provides additional stability. The square parts at the bottom of the picture have no clamping action, their sole purpose being to carry an indexing mark enabling the slots to be placed equally around the part.

*Photograph 11.5* A method of machining T-slots in a heavy duty faceplate.

## EXAMPLE 4

While this chapter is about mounting parts onto the worktable, in some cases it is often helpful to add some simple items onto the table onto which the workpiece is mounted; **Photograph 11.6** shows an example of this. I have particularly included this as it shows a type of clamp that is rarely used on the milling machine, very rarely in my case.

The casting is secured onto two square posts mounted on the table with the swivel pads on the ends of the clamp's screw being able to cope with the casting's taper. The flanges of the casting being already machined were firmly against the table's surface for this set-up and the top surface fully machined. The photograph shows the next stage, that is the machining of the two slots. For this, the round post was added to act as a reference for positioning the slots that had to be equally spaced about the V.

## EXAMPLE 5

If having to mount a round item onto the worktable for machining, many would consider using a V-block, but this example uses simple fences and has a number of advantages (see **Photograph 11.7**). Mainly, it can work with round parts from a few millimetre up, there theoretically being no

**Photograph 11.6** *Not a type of clamp normally used on the milling machine, but the swivel ends accommodate the casting's taper.*

**Photograph 11.7** *A method of securing a round item using two fences and an overhead clamp.*

upper limit. Other advantages are: much longer fences can be used for longer parts, it is more robust, being low set, and a V-block is not required.

To set up the method, fit the first fence using a square off the front edge of the table,

then, using a piece of packing as a spacer, position the second fence parallel with the first. The packing should be within one-half to three-quarters of the diameter of the part to be supported. Secure the second fence using a stud which can then also be used to secure

*Photographs 11.8* (above) and *11.9* Two examples of using the same method as that in *Photograph 11.7*.

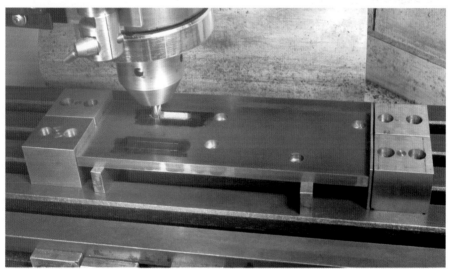

*Photograph 11.10* Shop-made versions of a split vise which permit much longer workpieces to be held than can be gripped in a conventional vise.

the workpiece. The photograph also shows the alternative method of using a V-block.

The method can also be set up on the angle plate if the part requires to be mounted vertically (see **Photograph 11.8**), and I have even used the method on the lathe when using a faceplate, so it is a well-tested method (see **Photograph 11.9**).

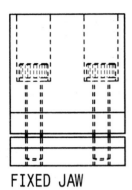
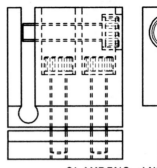
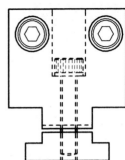

FIXED JAW

CLAMPING JAW

*Sketch 11.3 High profile clamps.*

## EXAMPLE 6

Split vises have separate fixed and moving jaws that are individually secured to the table and as a result much longer parts can be held. Also, the workpiece being placed directly onto the machine table, or on parallels, bypasses any errors that may exist in a vise if one were being used. With the last advantage in mind, their use is not limited to long parts only.

I do not possess a commercially available example of these, but **Photograph 11.10** shows my own shop-made version. The part being machined had to have two slots milled and the photograph shows that I used 'machine to' lines to avoid the need for measuring during the machining process, always worth considering.

Since having made these devices, the number of times I use a vise has dropped as I frequently use them, particularly for small parts (see **Photograph 11.11**). A major advantage of these is the total lack of jaw lift; also their small size would make them very convenient on the smaller milling machine. For the basis of the design, see **Sketch 11.3**.

*Photograph 11.11 The method in **Photograph 11.10** is not limited to large items.*

## EXAMPLE 7

Often it will be required to machine a surface totally, and if this is beyond the capacity of the available vise the workpiece will have to be mounted on the milling machine table. For this, clamps that just grip the side will have to be used, as seen in **Photograph 11.12**. When using this form of clamp, just fixed jaws will be required on the reverse side and a pair of your workpiece clamps will often be more than adequate. Do, though, bear in mind that the force generated by these clamps is quite large, so the fixed jaws will have to be very secure.

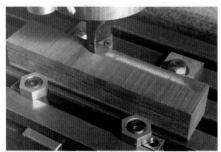

*Photograph 11.12 Low profile clamps are useful where a workpiece has to be machined fully over its top surface.*

As the clamps in the photograph are a commercial item, you may consider alternatively making some yourself; five alternative designs for these have been published in 'Model Engineers' Workshop Projects', Workshop Practice Series number 39. Also a very simple example was shown in **Photograph 8.6**, and as can be seen even copes easily with a part mounted at an angle, as do some of the others. **Sketch 11.4** shows the basis of this design.

Another method is to use two toolmaker's clamps as illustrated by **Photograph 11.13**. In this a piece of 1" strip is being reduced to 25mm wide to suit a metric design.

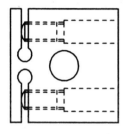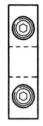

*Sketch 11.4 Low profile clamps.*

*Photograph 11.13 Two toolmaker's clamps are useful for securing some workpieces.*

# CHAPTER 12
# USING A DIVIDING HEAD

In this section we cover the subject of dividing heads. Unfortunately, being a complex subject, it is only possible to cover this relatively briefly, but the content should provide sufficient understanding of its use for the majority of home workshop tasks.

While there are other types of dividing heads, only the semi universal head is commonly available commercially (except for the universal head which is far too complex and costly for most workshops). Other heads that may be seen are frequently shop-made.

Input to the dividing head is set using dividing plates that have rings of holes at various pitches; the input is provided by the handle, then transferred to the output via a worm and worm wheel, having a ratio of 40:1. This results in a large number of divisions being available, but even with three dividing plates, each with six rows of holes, there are divisions that are not possible in the range above 50. The range available should, however, be more than adequate for almost all home workshops; just 125 and 127 are missing that do very occasionally have a use.

Typically if needing to achieve 35 divisions.This can be done using a ring with either 21 or 49 holes. Unfortunately, however, the manual with the dividing head does not quote alternatives which could be useful if a plate went missing. Let us work with the one quoted, that is 49 holes.

With the head having a ratio of 40:1, 40 turns of the input will result in one at the output. As the plate has 49 holes, the input will have passed 49 × 40 holes for one rotation at the output, making 1960 holes passed. If therefore the input was stopped at each hole, 1960 divisions would result, but

**Photographs 12.1, 12.2 and 12.3** *In these photographs the requirement was for one turn plus seven holes. To avoid the need to count the holes, the bars are set at a seven-hole spacing and moved after each division is completed.*

such a number is very unlikely to find a use. How then can this become a useful set-up? The answer is to stop the input after a given number of holes for each division.

If we consider every second hole, then the result will be 980, but if stopped every third hole it will give 653.33 divisions. Obviously, only when 1960 is divided by a whole number and giving a whole number is the resulting division a usable one.

Having the aim to achieve 35 divisions, we get $^{1960}/_{35}$, which equals 56 holes passed. With the plate having 49 holes this will be one rotation plus seven holes, an operation that would be fraught with possible errors, and therefore some form of assistance is needed. This is achieved by the arms seen in **Photograph 12.1** that have been set at seven holes' spacing and resting against the left side of the plunger. At this setting the first operation on the workplace is carried out. The plunger is then pulled clear, rotated one turn and a further seven holes as the arms indicate (see **Photograph 12.2**), and the second operation carried out. The arms are then rotated to rest against the left side of the plunger once more to indicate the position for the next division (see **Photograph 12.3**). The process is repeated until all 35 divisions have been completed.

Above, I mentioned that 35 divisions can also be achieved with a 21-hole dividing plate. Checking the mathematics, we get 21 × $^{40}/_{35}$ which equals 24. Being a whole number we know it is a workable division and will be achieved with one turn of the input plus three holes.

## NON-STANDARD PLATES

The dividing head will be supplied with a number of dividing plates, for which a chart will also be supplied giving the divisions possible with these. After some time the chart may be lost or additional dividing plates may be acquired; in these cases calculation will have to be resorted to.

To determine the number of holes in the plate to achieve the division required, use the formula:

$$W = \frac{R \times H}{D}$$

where D is the number of divisions required, R is the ratio of the dividing head, H is the number of holes on the plate, and W is a whole number and also the number of holes in the ring to be passed for one division. Knowing the head ratio and the number of divisions required, there are still two unknowns – how then can the answer be found? The main thing to remember is that W is a whole number.

Consider a ratio of 40:1 and requiring 55 divisions, the formula becomes:

$$W = \frac{40 \times H}{55}$$

Simplifying this by dividing both top and bottom by five, we get:

$$W = \frac{8 \times H}{11}$$

From this we see that W will be whole providing H is 11 or any multiple of this (22, 33, 44, etc.). This shows that a plate with any one of these numbers of holes will provide the required division.

Beyond this, additional reading should be sought, typically 'Dividing', Workshop Practice Series number 37, which also includes designs for shop-made dividing heads and various other methods of dividing, as well as extensive tables giving the divisions that are available with a wide range of division plates and dividing heads that have ratios of other than 40:1.

*Photograph 12.4 A simple application – milling a square to provide the drive when a handle is fitted.*

## MAKING DIVIDING PLATES

Making your own dividing plates for numbers not available with the supplied plates is not that difficult as, surprisingly, until you have done the mathematics that is, any error in a hole position is reduced by a factor equal to the ratio of the head. Typically, therefore, if a hole is one degree out on the plate the error will only be ¼₀th of a degree out at the head's output. If you need greater accuracy, then make a second plate using the first and the error will only be ¼₆₀₀th of a degree out. Various methods of making a plate are discussed in the book mentioned above.

## SOME EXAMPLES

The actual machining operations when using a dividing head are invariably very simple and need no explanation beyond setting it up.

*Photograph 12.5 Drilling Tommy bar holes around the outside of a tailstock die holder. The dividing head is a shop-made one.*

The following are therefore typical dividing set-ups.

*Photograph 12.6 When using a tailstock, long workpieces can be machined.*

**Photograph 12.4** shows a square being milled on the end of a shaft, its purpose being to accept a handle with a square hole so as to provide a secure drive. Frequently, a piece of workshop equipment will require Tommy bar holes to be drilled to enable it to be rotated using a bar or a C spanner. **Photograph 12.5** shows holes being drilled in a tailstock die holder, using a simple shop-made dividing head.

If long parts require dividing in some way, then the head's tailstock will need to be used as seen in **Photograph 12.6**, where a long shaft is having splines machined into its end while mounted between centers. As any free movement between the dividing head and the workpiece cannot be allowed, the drive arrangement is different to that used on the lathe for turning between centers and is shown in **Photograph 12.7**, which should make the set-up clear.

The photograph clearly indicates a feature of the semi universal dividing head that as yet

*Photograph 12.7 A special driving dog arrangement which is backlash free is required when working between centers.*

I have not mentioned: in addition to divisions set using the worm and wormwheel drive, a fixed plate is included with 24 holes spaced at 15°. This makes it easy to achieve the following divisions: 2, 3, 4, 6, 8 and 12.

Another rather more advanced operation, and one that most workshop owners will associate with a dividing head, is cutting

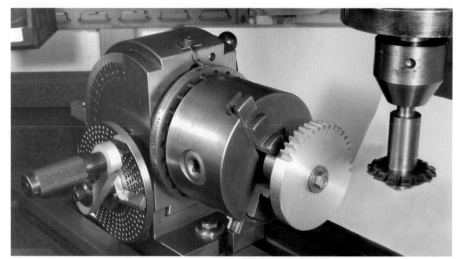

*Photograph 12.8 A common use for a dividing head is to machine a gear.*

gears, as **Photograph 12.8** shows. The machining operation is again simple; the complication is in the design of the gears themselves – what tooth form to use, for example. Also, as the space between adjacent teeth becomes greater as the number reduces, a separate cutter is theoretically required for each gear tooth number. The difference between adjacent gear tooth numbers is very small, however, and therefore one cutter can be used over a range. Even so, there are eight cutters required to cut the full range, that is 12 teeth to rack; five cutters will, though, give you the most likely range, that is 17–134 teeth. 'Gears and Gear Cutting', Workshop Practice Series number 17, is worth reading if more information is required.

## SETTING UP IN THE VERTICAL POSITION

While no task is being undertaken, **Photograph 12.9** shows that the semi universal head is capable of being angled through 90° and could be typically used in this position for drilling holes on a PCD. It

*Photograph 12.9 A semi universal dividing head can be set between 0° and 90°.*

could also be used as an alternative to a rotary table, though much less convenient, and its extra height may cause problems with available space between the table and cutter.

**Photograph 12.1** clearly showed (top left corner) that the head is calibrated for setting it at an angle; this, though, will not be exact and when set in the more normal horizontal position it should be set as **Photograph 12.10** shows after returning it from another angle. This should also be checked when the head is first purchased as it may not have been accurately set when manufactured.

## SPIN INDEXER

A spin indexer is an alternative, and much simpler, dividing device that is capable of being set at one-degree intervals. The 360 divisions are achieved by means of a plate that has 36 holes at 10° intervals and a short series of 10 holes placed at a spacing of 9° (see **Photograph 12.11**). To set a position, a pin is placed through one of the 36 holes and one of the 10 holes and working rather like the divisions of a vernier caliper achieves 360 one-degree divisions. Unfortunately, this still only gives 22 usable divisions, these being 2, 3, 4, 5, 6, 8, 9, 10, 12, 15, 18, 20, 24, 30, 36, 40, 45, 60, 72, 90, 120 and 180.

To set each position for the division required, it is necessary to convert these to angle rotated; typically for six divisions this will be 0, 60, 120, 180, 240 and 300 degrees.

As the plate is calibrated at each hole, that is 0, 10, 20, 30, etc., it is for the easier divisions just a case of lining up the pointer on the moving section and placing the pin through the first hole in the group of 10 and into the main plate. The divisions that can be achieved this way are 2, 3, 4, 6, 9, 12, 18 and 36.

For other divisions, the angle of movement will still have to be calculated and as these will not be divisible by 10 then the group of 10 holes will have to be used also.

Typically, if requiring 10 divisions, then this will be every 36°, being 0, 36, 72, 108, 144, etc. As the main plate is calibrated in 10° intervals it would be set just above 30 and with the pin in hole 6 rotated slightly until the pin lines up with the nearest hole in the main

*Photograph 12.10* After using the head at another angle, or when first purchased, the head should be set accurately upright, using the method shown, as the calibrations will not be that precise.

*Photograph 12.11* A spin indexer is a simpler dividing head giving just 22 usable divisions, but this does include the most common low numbers: 2, 3, 4, 6, 8, etc.

plate. For the next it would be just above 70 and using hole 2 followed by 100 and using hole 8 and so on.

As the method of securing the workpiece is using 5C collets, its use is rather limited. Because of this, I feel that to achieve the most from this device it would be beneficial to produce an adaptor to take the lathe's three jaw chuck and faceplate. This would consist of a solid 5C collet with the facility to support the chuck on the outer end. At the other end it could be threaded identically to the 5C, or just have a simpler tapped hole to take a draw bar.

# CHAPTER 13
# USING THE ROTARY TABLE

In this section we cover the subject of rotary tables, which in many workshops will be a more useful addition than a dividing head. Using one is much more varied and complex than using a dividing head, however, and is among the most complex of all milling operations done in the average home workshop. Because of this, the chapter is one of the largest in the book. To give the reader some insight into its use I will, as I did in the previous chapters, describe the processes by way of some actual examples of one being used.

## POSITIONING THE ROTARY TABLE
First, the rotary table must be positioned accurately in line with the machine's spindle, which is most easily done with a drilled center in the table's bore and a center in the machine's spindle (see **Photograph 13.1**). Rather than fixing the rotary table to the worktable and using the X and Y traverses to align it, place it loosely, engage the centers, then secure it. With that done, both the X and Y axis movements should be fully locked.

As it is probable that after doing this a cutter and cutter chuck will need to be fitted, positioning the table should be done with the machine spindle lowered fully; it can then be raised for adding the cutter and chuck.

**Photograph 13.1** *Centralizing the rotary table with the machine's spindle.*

Even so, it is possible that there may still be insufficient room to make the change. To overcome this problem, set one of the X axis stops at this position, enabling the table to be moved so as to obtain more room for fitting the chuck. With this done, it can be realigned with the machine spindle by returning to the stop and again secured.

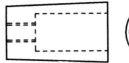

*Sketch 13.1* Taper to parallel adaptor. The thread is for jacking out the adaptor from the rotary table bore.

The rotary table will not come with a drilled center and this must be made. However, the bore through the center of a rotary table comes in two forms: a Morse taper or parallel. If your table has a taper bore, I would suggest that you make an adaptor consisting of a short length of taper, and, with a parallel bore, with this made other adaptors can all be made with parallel shanks making them easier to produce. The bore and the taper will have to be made without removing the part from the chuck as the two must be concentric. To avoid having to remove the rotary table from the machine table to remove the adaptor, make it with a threaded hole in its base allowing it to be jacked out from above (see **Sketch 13.1**).

*Sketch 13.2* Rotary table locator.

The basic form for the table locator is shown in **Sketch 13.2**, the dimensions of course depending on the table in which it is to be used. Concentricity is important, so the shank will have to be turned with a left-hand knife tool.

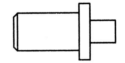

*Sketch 13.3* Workpiece locator.

With the table now on center, the part to be machined has to be placed; frequently this has a hole about which a radius has to be machined. To locate the part, produce another adaptor to fit in the parallel bore of the table and with a locating pin being a close fit in the part to be machined (see **Sketch 13.3 and Photograph 13.2**). The photograph also shows on the right the part about to be machined.

The workpiece can then be placed over the locating pin, clamped in place, and the required radius produced. This will be

## IMPORTANT!

A vital requirement when machining a workpiece in this way is that the rotary table must rotate in a way that opposes the rotation of the cutter. If the opposite is attempted the rotation of the cutter will attempt to take up the backlash in the table's feed mechanism. This is something that it will invariably do suddenly and with disastrous results, such as a damaged workpiece and/or a broken cutter. This is discussed in more detail in Chapter 16. However, if you are using a slot drill to machine a slot, then the table can turn either way.

*Photograph 13.2 An adaptor for centralizing the workpiece with the rotary table. The workpiece to be machined is seen on the right.*

done by traversing the milling machine table by the radius required plus half the diameter of the cutter being used; the cutter is then lowered and the radius made (see **Photograph 13.3**). A thin piece of card below the workpiece will enable the cutter to be lowered sufficiently to machine the part fully without making contact with the adaptor. The photograph shows the remnants of the card used. If too much needs to be removed at one pass, then go beyond that setting and work back towards it.

    **Photograph 13.4** shows the ends of a con rod being machined in much the same way, but this also required a much larger radius to be machined on its arm. For this, a fixture was made to locate the part correctly so that the larger radius could be machined, **Photograph 13.5**. The two pins locate the part, the nearest large hole is for clamping it and the farthest is placed over the central locating pin so that the correct radius is

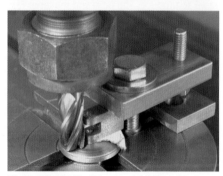

*Photograph 13.3 Machining the workpiece seen in Photograph 13.2.*

*Photograph 13.4 A similar operation to that in Photograph 13.3, this time for a simple con rod.*

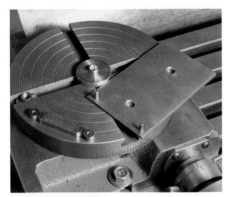

*Photograph 13.5 A jig for positioning the con rod so that a large radius can be machined onto its arms.*

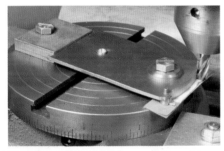

*Photograph 13.6 Machining the con rod's arms using the jig.*

*Photograph 13.7 The part finished con rod.*

achieved. **Photograph 13.6** shows it being used and **Photograph 13.7** the result.

In some cases, a workpiece will need securing with a central screw rather than overhead clamps, as were used in the examples above, and for this the locator will need to be captive prior to fitting the part to be machined. A method of achieving this is illustrated in **Sketch 13.4** showing that, while it is quite simple, it has more parts than the single one shown in **Sketch 13.3**. The assembly consists of three parts, some of which will need to be made in a number of sizes.

The smaller diameter of part A is a close fit in the table's bore, also having a countersunk head that permits it to be used for centralizing the rotary table, and must therefore be concentric with the smaller diameter, as must the outer diameter of the flange. Two thread sizes will probably suffice, M3 for the workpieces having bores up to 6mm and M6 for those larger.

Just one part B is required, the only important features being that the larger bore must be a close fit on the flange of part A and both bore diameters must be concentric with one another.

As part C is used to locate the workpiece, one will be required for each bore diameter to be located. The larger diameter, as drawn, must be a close fit in part B; in some cases, though, the locating diameter will be the larger where large bore diameters are being located. The two diameters must be concentric. Additionally, four T-nuts will be required to secure the assembly. The parts are shown in **Photograph 13.8**; unfortunately this device is not available commercially so it does have to be shop-made. The photograph was taken using a mirror so that the rear of part B could be seen and shows the anti turn pin that prevents part A from turning when the workpiece is being secured.

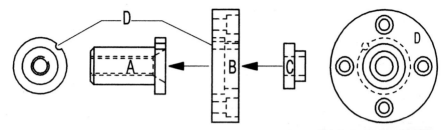

*Sketch 13.4* Workpiece positioning jig. (A) Rotary table and upper plate locator. (B) Upper plate. (C) Workpiece locator. (D) Anti turn pin and notch.

**Photograph 13.8** *This kit of parts enables workpieces to be located, then secured with a central screw. The photograph was taken with a mirror to allow the rear of the main part to be seen. See text for more details.*

In use, part A will be placed into the bore of the rotary table and the table centralized, as in **Photograph 13.1**, and then secured on the machine table. Part B is then placed over this and secured to the rotary table using four screws and T-nuts. Part C is then placed into the bore in part B and the workpiece placed over this and secured with a screw into part A. **Photograph 13.9** shows a very small part being machined using the method, and **Photograph 13.10** a much larger one. Note the difference in the size of the securing screw being used, achieved by using an alternative part A.

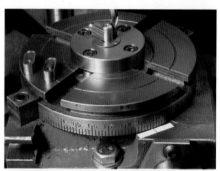

**Photographs 13.9 and 13.10** *The locating assembly being used for a small part (13.9) and a much larger part (13.10).*

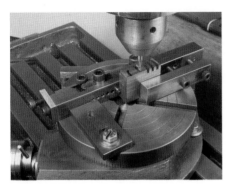

**Photograph 13.11** *Machining soft jaws for use in a three jaw chuck.*

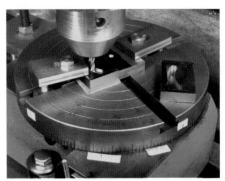

**Photograph 13.12** *Machining a closed end slot.*

**Photograph 13.11** is of a much more complex task, showing soft jaws for a three jaw chuck being made, and is included to emphasize the benefit of using a fixture for some operations. Space does not permit a full explanation, but I will attempt briefly to provide the basics for any reader who may like to produce jaws for their chuck.

The groove has to be milled in two passes as the tooth's internal and external radii differ. Achieve an initial estimate of the radius required by offering up some round items, typically paint tins, to the chuck's normal jaws, and then produce a cardboard template to check your considered values. The values are not crucial, but any error should be with the outer radius having a minus value (more curved) and the inner radius a plus value (flatter). Note also that the peak of the tooth is not central across the width of the jaw but a rule dimension will be adequate.

The jig is drilled to take stop pins at a spacing equal to the jaw's pitch. Three pins are then required, one to be used for each jaw, but while their diameters are not important one must be smaller than the larger by one-third of the jaw's pitch and the smaller by two-thirds. I realize that this is a complex example, but I have included it to illustrate the versatility of the rotary table.

## CLOSED END SLOTS

Unfortunately, most rotary tables are not fitted with rotational stops, so one must work to its calibrations; because of this, marking them with tabs will help to avoid going too far due to a loss of concentration (see **Photograph 13.12**). Notice how 'machine to' lines have been scribed onto the workpiece to enable the tabs to be easily positioned on its first traverse.

This example also illustrates another method of locating a workpiece where the center of the curve being machined is outside its edges. In this case, a fence has a drilled center added such that if this is centralized on the table and the workpiece placed against it the slot will be correctly placed (see **Photograph 13.13**). Assuming the original part drawing was dimensioned to the center of the slot, the table will be traversed just by the value of the radius as the diameter of the cutter does not need to be taken account of, as it would for an external radius. If you refer to **Photograph 13.12** you will see that the workpiece is also aligned with the end of the fence to locate it endways. As there were two parts, the method ensured that they were accurate and both the same.

## FITTING A CHUCK OR FACEPLATE

Parts are not always mounted directly onto the rotary table, but use a chuck or faceplate, itself mounted onto the table. **Photograph 13.14** shows a threaded adaptor on which they would be fitted, and again the rotary table would first be placed central with the machine spindle, then the chuck adaptor, also having a drilled center, would be loosely placed, centralized, and finally secured on the rotary table. The chuck can then be fitted with the knowledge that it is both central to the milling machine spindle and the rotary table.

However, there is a potential problem when using a chuck in this way and it is essential that it is not overlooked. If machining the outer face of a workpiece fitted into the chuck, say to machine a hexagon, the action of the cutter would be attempting to unscrew it from the adaptor; it should not therefore be used in this way. Its use would typically be to place holes on a pitch circle diameter, using the chuck for a small item or the faceplate for something larger. A major advantage of using a chuck or faceplate to hold the workpiece is that it can be machined on the lathe and then transferred to the rotary table with the part still accurately central.

An alternative method of securing a chuck is to have a backing plate fitted to it that can then be secured directly to the rotary table (see **Photograph 13.15**). If, however, the workpiece is being transferred already fitted in the chuck, then it will require a drilled center for locating the assembly. If this is not possible, the part will have to be removed and the adaptor seen in **Photograph 13.2** fitted into the chuck for it to be centralized. The workpiece is then returned. Concentricity cannot be guaranteed, but the error should be very small and of no consequence in most cases.

Photograph 13.13 As the center of the radius of the part in Photograph 13.12 was outside its edge, a locating fence had to be positioned using a center.

Photograph 13.14 An adaptor for fitting a chuck or faceplate. Note this also has a drilled center for positioning it centrally on the rotary table.

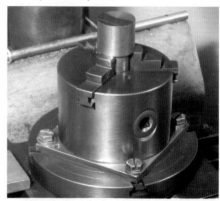

Photograph 13.15 An alternative and more secure method of fitting a chuck to the rotary table.

**Photograph 13.16** shows a hole being cut in the backing plate for the chuck and illustrates another use for the rotary table, that is cutting large round holes. Note how the plate is raised on packers so that the cutter can break through without damaging the table's surface. The same method could be use to cut a disk rather than a hole.

**Photograph 13.16** *The rotary table can be used to cut a hole or to produce a disk.*

## USED VERTICALLY

It is likely that most uses for the rotary table will be with the table horizontal, but occasionally the task in hand will require it to be vertical. Another occasional feature is for the workpiece to be supported by a tailstock. **Photograph 13.17** shows both situations.

## USED AS AN ADJUSTABLE ANGLE PLATE

A less obvious use for the rotary table is to set it up as an adjustable angle plate. With an angle plate fitted to the rotary table the plate is initially set vertical (see **Photograph 13.18**), then the dial is zeroed and the angle plate rotated by the required amount (see **Photograph 13.19**). While not showing an actual application, the photographs should make the principle clear.

**Photograph 13.17** *The rotary table in use in its vertical position and with a tailstock supporting a long workpiece.*

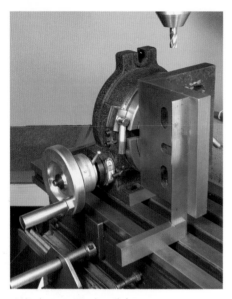 

*Photographs 13.18 and 13.19 In the absence of a tilting angle plate, using the rotary table with a fixed angle plate may provide the facility.*

## USED AS A DIVIDING HEAD

A few larger rotary tables can have dividing plates added, so that more precise angular movement can be achieved, resulting in almost all tasks that could be done using a dividing head then being able to be carried out using a rotary table. In this case, the details in Chapter 12 would largely apply to a rotary table (see also **Photograph 6.7**).

A rotary table is capable of a very wide range of machining tasks and those illustrated in the chapter are but a few. They should, though, provide the reader with sufficient knowledge of the subject to cope with all but the most complex requirements.

A final tip. As the required machining is largely carried out using the handwheel on the rotary table, I still find myself instinctively using one of the milling machine handwheels. Fortunately, they are all easily removable, so I remove them, which avoids me using them and losing position which can then be difficult to re-establish precisely. If your milling machine handles are not easily removed, then place a plastic bag over them.

CHAPTER 14
# USING OTHER ACCESSORIES

## 5C COLLET ADAPTORS

The provisions for holding round components for milling are very limited, with these devices being the only ones commonly available commercially (see **Photograph 6.9**). The collets are a one size only device; typically a 1" collet (25.4mm) cannot (or should not) be used for holding 25mm diameter components. You could of course use the method I proposed as it is much more adaptable (see **Photographs 11.7, 11.8 and 11.9**).

Using the adaptors is very straightforward, the only consideration being how deep the workpiece should enter the collet. If only a very light machining task, then a depth equal to half the diameter of the part being held should suffice, but for something more demanding a depth equal to the parts diameter should be considered a minimum.

Both can be mounted with the workpiece horizontal or vertical, but that on the left has the advantage of having a dead length mechanism. With this, rather than the collet being pushed, or pulled, into the taper, the collet is held stationary and the taper moves forward to close the collet. This avoids problems with the front of the collet finishing up at different positions due to the items being held having slightly differing diameters. This is useful for second operation tasks where precision is required, but is rarely a requirement in the home workshop.

For large diameters, a Keats angle plate, more associated with the lathe, could be used on the milling machine (see **Photograph 14.1**).

*Photograph 14.1 Using a Keats angle plate, usually associated with the lathe, on the milling machine to hold a large-diameter workpiece for machining.*

## BORING HEADS

The purpose of a boring head is to produce holes on the milling machine, larger than can be drilled and where the component is too large to be accommodated on the lathe's faceplate for the purpose. The heads are provided with adjustment (see **Photograph 14.2**) for the size of the hole being bored, but this is quite limited, say 20mm (approximately ¾"). However, this is overcome by having a number of holes into which the cutter can be placed (see **Photograph 14.3**). With a large hole being bored, the diameter would be increased as far as the adjustment allows, then the adjustment moved back and the cutter moved up one hole. As shown, the head in the photograph will produce holes up to 75mm (3"); for even larger holes, the cutter can project from the side of the head (see **Photograph 14.4**).

The process is quite slow, as the depth of cut and feed rate will have to be kept to a

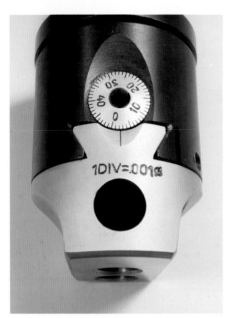

**Photograph 14.2** *A boring head is provided with micrometer adjustment for setting the diameter being bored.*

minimum due to the length of the cutter being used. This will result in a largish number of traverses – typically, to go from a 16mm drilled hole up to say a 40mm diameter final bore, 24 passes if the diameter is increased at a rate of 1mm per pass. With experience of the method, and your machine, you may find it possible to increase the diameter by larger amounts per pass. **Photograph 14.5** shows an example of a boring head being used.

If machining a blind hole, or to a step, then it is essential that the down feed stop is set. This avoids the long cutter attempting to take a wide cut if attempting to go too deep. Also, unlike boring a hole on the lathe where the cross slide could be used to face the bottom of the hole, this cannot be done with a standard boring head. Controlling the depth using the down feed stop will go some way to achieving an acceptable finish.

**Photograph 14.3** *As the traverse on a boring head is quite small, more than one position for the cutter is provided to increase the range.*

Fine feed on a few small milling machines is achieved by moving the head up and down its round column which can allow the head to swing very slightly side to side as it cannot be clamped when using a boring head. I have overcome this by using the drilling down feed; it does, though, need to be done with considerable care to avoid feeding the cutter too fast.

**Photograph 14.4** *For large diameters, the tool can be mounted from the side.*

**Photograph 14.6** *Using a boring head as a fly cutter, in this case as no other cutter had the reach required.*

## AS A FLY CUTTER

Another use for a boring head is as a fly cutter. In this case, the tool is set to machine at a fixed diameter, suitable for the task. **Photograph 14.6** shows a rather extreme example in which it is being used to machine into a V as no other available cutter had the reach required; the method worked well, even with the tool being so long. The diameter that it was cutting was set quite small, probably 16mm (⅝"), so that it was very much like using a single tipped end mill of that diameter, not that unusual therefore. Of course, if used normally as a fly cutter at a larger diameter, one of its shorter tools would be used.

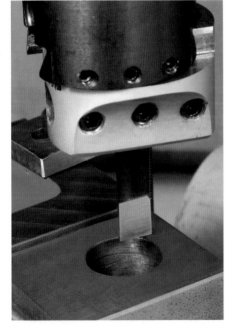

**Photograph 14.5** *A boring head in use.*

I know I have mentioned it before but I want to emphasize how beneficial it is when milling to work to 'machine to' lines wherever possible. These, setting the width of the V being machined, should be visible in the photograph.

# CHAPTER 15
# POSITIONING THE WORKPIECE

In the previous chapters, the various methods of securing the workpiece for machining have been discussed. With them mounted, it now has to be placed so that the correct amount of metal can be removed by using the leadscrew dials. Obviously, this can only be done if the relationship of the cutter to the workpiece is known initially; frequently this will be an edge on the workpiece relative to the machine's spindle. There are a number of ways of doing this, some very simple and others less so.

## TESTING TO AN EDGE

The simplest method, needing no special equipment, is to bring the rotating cutter slowly towards the workpiece edge until it becomes apparent that it has touched the part. This may be visually by spotting the first sign of the face being machined, audibly, or more likely a combination of the two. The visual element can be improved by marking the workpiece with marking blue; also, to improve the accuracy of the result, carrying out the test again adjacent to the first may be worthwhile. To do this, note the dial reading obtained and bring the cutter to within say 0.05mm (.002"), but this time, at this point, traverse the table much more slowly.

A variation of this method that has been used for many years is to take a piece of very thin paper, such as cigarette paper, moisten it so that it adheres to the workpiece side and traverse the table as above. When the piece of paper is pulled from the part, then the cutter can be considered as touching it. This has the advantage that the workpiece is not marked if the surface being used is not to be machined in any way.

A major advantage of the above two methods is that they can be performed with the cutter that will be used. Those that follow use a special device that will have to be removed to allow the cutter to be added; maybe also the collet in the chuck will have to be changed.

## ELECTRONIC EDGE FINDER

These are available from many sources and come with a variety of names, such as 'touch point sensor' and 'edge finder', but all

**Photograph 15.1** *Using an electronic edge finder to locate the edge of a component relative to the machine's spindle.*

work outwardly in the same way – a sensor touches the workpiece and a bulb or LED lights. They are called electronic, but some are just a torch bulb and a battery; no doubt some do have electronics to control the lamp. If so, I would anticipate that these would be more sensitive and therefore less affected by contamination on the workpiece surface, but I have no experience of this. Some do include a buzzer as well as a lamp.

In use, the finder is placed in the cutter chuck and with the spindle stationary the workpiece is slowly advanced until it contacts the finder, when the lamp will light (see **Photograph 15.1**). In view of its length, it will not be that difficult to deflect the finder slightly; therefore, note the dial reading, wind the table back a little and advance the table once more, this time much more slowly.

If the sensor is then raised above the workpiece, and the table traversed by half its diameter, the machine spindle will be exactly above the workpiece's edge, or at least that is the theory. The sensor may not, however, be running true, so a second test is required. To do this, turn the machine spindle through 180° and repeat the process; the machine spindle center will be the mid-point between the two readings taken. If you find that the test only shows up a very small error and this is repetitive during a few more uses, then you may consider that a single test is all that is needed except for very critical applications.

Very often, using the end mill will be more than adequate as the complications of removing the electronic finder and adding the cutter would rule it out. However, there will be instances where machining has to take place relative to some aspect of the workpiece that the end mill cannot access, a narrow channel for example. In this case the smaller finder, likely to be around 5mm (.20") diameter, will have to be used, or some other method found.

Another use for it is to locate the machine spindle centrally above a hole in the workpiece. To do this, you will need to know the diameter of the hole as, due to the backlash in the leadscrews, using it to determine both sides of the hole will be difficult. Set the finder centrally across one axis – a rule dimension will be OK – and find the side of the hole in the other axis.

Now subtract half the finder's diameter from half the hole's diameter and from this calculate the reading of the leadscrew dial for the hole's central position. However, backlash will result in an error if you just wind the table back to the reading required. To overcome this, wind the table back past the required reading and then forward again to the value required. Having found the central position on one axis, lock the table at this and repeat the process for the second axis to establish the center of the hole. This can be seen to be an awkward method and I consider that using a DTI as suggested later in the chapter would be preferable.

## WIGGLER

This is a quite different device, being totally mechanical and consisting of a number of different items (see **Photograph 15.2**). I

*Photograph 15.2 The set of parts that make up a wiggler.*

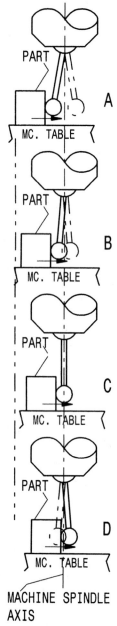

will deal first with the most frequently used part. I should add, though, that there are others that look rather different to those in the photograph but all work on a similar principle; that in the photograph is I think the most common.

As an edge finder, use the bar that has a ball at either end with the larger fitted into the holder, which is tightened such that the arm will move easily but just tight enough for it to hold its position. It is then placed in the drill or cutter chuck and deliberately set off center so that it swings in a circle when the machine is started up and the workpiece brought up to the finder (see **Sketch 15.1A**). If the workpiece is continually moved the circle the finder is making will diminish (see **Sketch 15.1B**), until it eventually runs true (see **Sketch 15.1C**). If then the workpiece is fed further, it will attempt to run in a circle again, but part of this theoretical circle will be within the workpiece (see **Sketch 15.1D**), which of course cannot happen and the arm will swing rapidly to one side at the point at which it attempts to go past center.

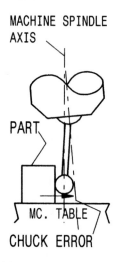

*Sketch 15.1 Wiggler as edge finder.*

*Sketch 15.2 Chuck held wiggler.*

The advantage of this method is that it is unaffected by any inaccuracies in the chuck holding the device as **Sketch 15.2** attempts to display; compare this with **Sketch 15.1C**. My experience with using the device is that the result is not totally repeatable, giving one reading for one test and then something different if the test is repeated; the difference is, however, small. Perhaps I should experiment with making the ball joint tighter or looser to determine the effect.

### TESTING TO SCRIBED LINES

Sometimes a workpiece will be marked out using scribed lines on a blue surface, most often for drilling but sometimes for other reasons. Aligning the machine spindle with the scribed line is another task that the wiggler can perform. In this case the ball ended tester is replaced by one with a needle point, but this again will be off center and needs to be set using a similar method to that used with the ball ended device. This is done using a suitable face on the workpiece which by traversing the table forces the pointer into smaller and even smaller circles until it appears to be running true, at which point the table is stopped. If the controlling face is moved too far, then the pointer will swing to one side and the pointer will have to be reset.

Now having the pointer on line with the machine's spindle, it can be used to align the spindle with the mark, or markings if crossed lines, on the part to be machined (see **Photograph 15.3**). Using a magnifying glass will help to improve the accuracy of the result. Personally, I find this the most useful aspect of using a wiggler, as I find the cutter against the workpiece edge to be adequate, and much quicker, for almost all requirements.

**Photograph 15.3** *Using the wiggler needle point to set the machine's spindle to a scribed line on the workpiece.*

### LASER CENTER FINDER

An alternative, and more modern, method is to use a laser center finder. but unlike the wiggler it has to be set up to compensate for errors in the chuck holding it or inherent in the device itself. To do this, the machine spindle is rotated by hand and the circle it scribes observed. This is then corrected by adjusting four screws within the device, much like setting a four jaw chuck until the point remains stationary. My impression is that this process would take longer than setting a wiggler to run true, but of course it depends of the level of accuracy being sought. Even when set, it would only be precise if the finder and workpiece are at the same distance apart for each test being carried out.

I do accept that if your machine, chuck and collets are precision items then perhaps my comments are more theoretical than actual. I should add that the laser device can be used for edge finding by testing the point at which the beam just drops off the top surface as the workpiece is traversed.

Neither method is therefore perfect, but the wiggler does have one other use that it performs better than the laser, as it actually indicates the error in the setting being carried out – that is to align the machine spindle with a round hole or raised boss.

**Photograph 15.4** *Using the wiggler's DTI attachment to set the machine's spindle central with a component's bore.*

## TESTING TO A VERTICAL CURVED SURFACE

To do this, another attachment is used in the wiggler that permits a dial test indicator (DTI) to be used. In this case, unlike using the ball end or pointed attachment, the collet that holds the ball is tightened fully to prevent the indicator from moving. The machine spindle is rotated by hand and the readings on the indicator noted, and from these the amount that the table needs traversing can be established; note that this will be half the total indicator reading. The method can be used both on internal (see **Photograph 15.4**) and external surfaces. In this test, the method automatically compensates for errors in the spindle and chuck being used, and as a result there is no setting-up procedure to be carried out.

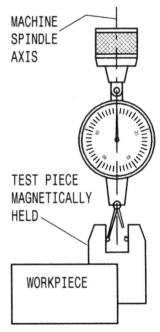

**Sketch 15.3** *Edge finding method.*

As a DTI has a limited range, the method is best set up initially just above the circle to be tested, doing this just visually. When close, the indicator can then be lowered to contact the workpiece and checked one axis at a time. With the error noted, the table is traversed by half the error and once more tested, repeating this until correct, when the second axis can be tested similarly.

Having introduced using a DTI with the wiggler, I will return to edge finding and using this with a simple device attached magnetically to the workpiece as illustrated by **Sketch 15.3**. In use, the spindle is rotated and the difference in the reading between each side noted. The table is then traversed by half of the error and the test repeated. When both sides give the same reading, the spindle is exactly in line with the edge of workpiece. Again the method compensates for errors in the spindle and chuck being used.

**REQUIRED?**

I would suggest that the workshop owner, at least at first, uses the method of just bringing the cutter up to the workpiece edge, with or without a piece of thin paper, as this I consider will be adequate for the vast majority of tasks involved. It is also quick. It will not, of course, compensate for the cutter not running perfectly true, but if using a good-quality cutter chuck the error should be minimal.

I would also obtain the wiggler, both for aligning the spindle accurately with marked lines on a workpiece using the needle point and to use it with a DTI for locating round holes or bosses. These tasks I consider it will perform easier than using a laser finder. This is my considered opinion rather than based on using a laser device.

## CHAPTER 16
# MAKING THE FIRST CUT

We are nearing the end of the book and yet the subject of cutting metal has not been aired. This is deliberate, as I wished to emphasize that using the milling machine is very much about set-up and far less about carrying out the machining operation, much of which is relatively simple and repetitive. However, there are some basic principles that must be taken note of – typically, the effect of the feed direction relative to the cutter rotation.

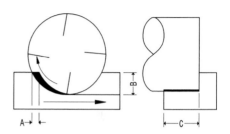

*Sketch 16.1* Comparison between depth and actual width of cut. (A) Actual width of cut, maximum, much exaggerated. (B) Apparent width of cut. (C) Depth of cut.

### WHICH CUTTER?

Having set up the component for machining, the first consideration is which type and size of cutter to use. First, I will deal with the common HSS end mills and then discuss the pros and cons of using the more recent replaceable tip tools. Please excuse me if there is some repetition of comments I have already made, but some things tend to fit into more than one aspect of the book's content.

### END MILLS: HIGH SPEED STEEL (HSS) TYPES

The advantage of the basic HSS end mill is that it has a very sharp cutting edge, made possible by a generous helix angle, and because HSS can be sharpened to a much sharper edge than can carbide. These therefore are very much more appropriate to use on most home workshop machines, particularly the smaller ones. However, if you are machining an iron casting, then these may contain the occasional very hard spot that will easily ruin your HSS cutter, so using a carbide tipped tool is almost essential. Alternatively, reserve an old, somewhat blunt, HSS cutter for the purpose; usually this will

be acceptable but once in while it will be ruined. If of course you are just machining a plain surface, a fly cutter could be used, as these can be easily sharpened should the cutting edge be damaged (more about these later in the chapter).

While called an end mill, almost all of the work is done by the cutting edges on its side, therefore the depth of cut is much more responsible for the load on the machine than is the width. This is best understood by reference to **Sketch 16.1**. Typically, doubling the depth will double the load but increasing the width will only increase the duration for which the cut is being taken, and therefore the power consumed. I say 'only', but more power will mean more heat generated, so care must be taken not to overheat the cutter as this will lead to its cutting edge being prematurely dulled.

If your machine seems unhappy with the amount you are asking it to remove, reduce the depth of cut and/or the feed rate, as reducing the width will have very little effect. Even so, there is a limit to the width of cut that should be taken – that is, slightly less than half the cutter's diameter, say 40% maximum. If

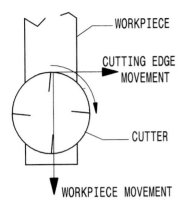

Labels: WORKPIECE, CUTTING EDGE MOVEMENT, CUTTER, WORKPIECE MOVEMENT

**Sketch 16.2** *Relationship between cutter and workpiece movements.*

you go beyond this point and past 50%, then you will find that the load on the leadscrew being used to traverse the table will increase appreciably. My own explanation for this is that as the cutting edge approaches and passes 50% the cutting edge and workpiece traverse are moving at 90° to each other, only possible if the cutter is deflected (see **Sketch 16.2**). If, however, you are using almost the full width of the cutter to surface the workpiece using just a very shallow cut, say 0.2mm (.008") deep, then this should be acceptable.

Initially, therefore, I would suggest for a step being cut a depth of no more than one-third and a width of no more than ⁴⁄₁₀ths of the cutter's diameter. With experience and a reasonably robust machine, a large mill drill typically, or bigger, and a sharp end mill, you should be able to increase the depth.

With regard to the diameter of the cutter, use the largest one that your cutter chuck can hold as it will be more rigid and can remove metal more rapidly. If the step being machined is wider than can be produced at a single pass, second and subsequent passes still follow the same rules.

If machining an open ended slot, then use a cutter equal to the width of the slot being made, but if this is not available use the next

smallest and widen the slot with a second pass. This would seem to be contrary to my comments above regarding using no more than ⁴⁄₁₀ths of the width of the cutter. In this case, though, there is no alternative, but the depth should be reduced to a maximum of ⅛th of the cutter's diameter; the feed rate will also need to be kept well down.

### END MILLS: CARBIDE TIPPED

If you have a tipped end mill with the helical form tips, then these can be used in place of the HSS end mills, but at the smallest sizes they are likely only to have a single tip; larger sizes will have two. This compares to four cutting edges with an HSS end mill, so unless your machine is robust enough to run them at the high speeds that they prefer the HSS end mill is to be preferred.

### SLOT DRILLS

If you are machining a recess or an enclosed slot, the most common end mills are not capable of doing this easily. This is due to the fact that they can only be plunged into the workpiece by a very small amount as they are not center cutting. Normally, for this operation, therefore, you will use a slot drill, as these, having one cutting edge that passes center, can be plunged in the same way as a drill. Having plunged the cutter, you can then traverse it to produce a slot or the four sides of a recess, repeating the process until the required depth is achieved. When using these, I would suggest a maximum plunge depth of ⅛th of the cutter's diameter, and, having only two cutting edges, again the feed rate should be kept well down.

If you do possess center cutting end mills as mentioned in Chapter 2, then the feed rate can be increased, but still limit the depth of the cut to a maximum of ⅛th of the cutter's diameter, this really being the same as milling an open ended slot as mentioned above.

## INTERCHANGEABLE TIP CUTTERS

I have referred above to helical form tipped end mills, but there are many more with non-helical tips that are only suitable for surfacing, at least on the average home workshop machine. When using them for this purpose, keep the depth of cut down to say 0.2–0.5mm (.008 — .20") deep. If your machine will not permit them to be used at the speed and feed rates they are capable of, then their use should be reserved for machining just cast iron castings.

### OTHER SPECIALIZED CUTTERS

These come in a number of forms, but each one has a specific task to perform. With few options, such as HSS or tipped, choice is simple (see **Photograph 2.5**).

### BALL NOSE END MILL

In the average home workshop, the use of these will be to produce a fillet between two surfaces at 90° to each other, and they can be used in the same way as a conventional end mill. However, if there is a substantial amount of metal to remove, then remove the bulk using an end mill as this has four cutting edges, leaving a square fillet which can then be removed using the ball nose tool. Channels, enclosed or open ended, can also be produced, the process being identical to using a slot drill.

### ROUND OVER CUTTERS

These produce an external radius between two surfaces at 90° to each other, and having four cutting edges are used much like an end mill.

### DOVETAIL CUTTERS

Dovetail cutters are used for milling machine slides and are made in a range of angles, but 60° is the most common. When using a dovetail cutter, the bulk of the metal should be removed using a normal end mill and then the angled faces produced using the dovetail cutter. If you are machining a deep dovetail, the cutter will be machining a wide face and vibration may result on smaller machines. This can be partially overcome by initially making it just half depth and just short of the width required, followed by repeating the process for the lower half. Then finish the dovetail by machining the complete face at one pass to avoid the possibility of there being a step in the surface.

When machining dovetail slides, it is most important that the two sides of the inner part are parallel, so ideally both will be machined without removing the part from the machine table. If the capacity of your milling machine will not permit this, then you will have to very carefully reposition the part for the second side. The outer part is less important as the gib strip will compensate for any error.

Measuring the part to determine that the position of the angled face is correct can be a problem but one that can easily be overcome by using short lengths of rod from which measurements can be taken, as **Sketch 16.3** shows; the outer part can be measured similarly. If you need help with the mathematics, this was covered in the 'Metalworker's Data Book', Workshop Practice Series number 42. The information

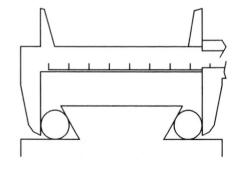

*Sketch 16.3 Measuring dovetails.*

is also found in 'Machinery's Handbook,' as is just about any other reference material required in the home shop.

## T-SLOT CUTTERS

In the home workshop, these are likely to be used to cut T-slots for such items as a rotary table or an angle plate. After having made the leg to almost the full depth of the T-slot, say -0.2mm, using a standard end mill, the cutter is used to cut both arms fully at one pass and at the same time finally finishing its base. However, the depth of the arm can be increased by lowering or raising the cutter and making a second pass.

## SLITTING SAWS

To enable these to be used on a vertical milling machine a mandrel will be required, either having a taper shank for the machine's spindle or a parallel one to be used in the machine's chuck.

As explained below in more detail, your machine spindle is likely to be very slightly off perpendicular, resulting in the slitting saw not being parallel to the table's traverse. The result of this is that the slot being cut will be attempting to follow the plane in which the saw sits rather than the traverse of the workpiece, and the resulting slot will probably wander. This is more likely with thinner and larger diameter cutters attempting to cut a deep slit at one pass. Unless otherwise dictated by the component being made, do therefore select the thickest saw available, as it will be less likely to wander.

Whatever the thickness of the saw being used, the following method is recommended, although it is much more important when thin deep slots are being made. Make the first cut just 2mm (.08") deep and check the result; then follow similarly with second and third cuts of 2mm. Use a very gentle feed so that you are not placing too great a load onto

**Photograph 16.1** *A deep slot being cut with a relatively wide slitting saw.*

the cutter, which would encourage it to flex. With the cut now at 6mm (¼") deep, deeper cuts can be taken. With thicker cutters, say 3mm (⅛") plus, there is less need to be this cautious, so you could commence with cuts of around 5mm (.20").

One other situation that it is easy to overlook is that, as you are working with a large diameter, the machine speed must be kept well down. If you do not do this, then the cutter will rapidly overheat due to there being very little metal in the cutter to absorb the heat generated, and the cutter can be easily ruined in a matter of seconds, especially if a very thin cutter is being used (there is more about cutter speeds later in the chapter). However, if you have the luxury of a flow of coolant to cool the cutter, it will be less of a problem.

**Photograph 16.1** shows a part being slit almost in two; this is to permit the leaf to be slightly flexible. Note how the workpiece is additionally supported by the angle plate to ensure that the part is not moved within the vise's jaws. While not seen here, an item of similar thickness is placed in the vise at the other end of the jaws. Being a deep and wide

**Photograph 16.2** *A fine slitting saw making a cut in a split bearing.*

**Photograph 16.3** *Fly cutters are useful for machining larger surfaces at one pass.*

slot, this is definitely a situation that benefits from a thicker cutter.

**Photograph 16.2** shows a simpler application that can be done using a much thinner saw, and as can be seen the task is being carried out on the lathe. Even when you have a milling machine, do not lose sight of the fact that in some cases milling in the lathe will be an easier or quicker option.

### GEAR CUTTERS

These are used in conjunction with a dividing head of some form to cut a gear, as seen in **Photograph 12.8**. Like the slitting saws, they are mounted onto a mandrel and again the speed should be kept down, although with them being more bulky overheating will be less likely. If also the gear you are cutting has large spaces between each tooth, then it may be advisable to make them in two passes, say cutting all the teeth at reduced depth then going round a second time to complete the gear.

### FLY CUTTERS

These in the past were commonplace in the home workshop, due no doubt to the economics of using one compared to a commercial cutter, and were extensively used to machine large faces of workpieces held on the lathe's cross slide. Their use on the milling machine now is less common, but still worth considering for large surfaces.

In use, a fly cutter has some special requirements due to it working at a larger diameter and having just one cutting edge. Firstly, requiring a low speed. Assuming the cutter is working at a diameter of 48mm (1.89"), then, using the guide lines I include later in the chapter, a speed of 125rpm will be required. Secondly, to achieve the same feed per cutting edge compared to a standard end mill having four cutting edges the feed rates will have to be reduced to 25%.

Success when using a fly cutter is largely achieved by ensuring that the shape of the cutting edge is both correct and finely honed. With that being met, my reservations above regarding feed rate are over-cautious, and the feed rate can be increased; experience will establish appropriate values.

Fly cutters as purchased frequently need the cutter bit to be both shaped and sharpened, and initially this should be done so as to emulate the cutting tip angles of a standard end mill, particularly the helix angle. If this is done, however, with such an acute corner, the machined surface will not be that smooth, as **Sketch 16.4A** shows. The vital requirement is that the tip should then be honed to have a very slight curve on its lower face, the width of which needs only to be a little more than the feed rate per tooth, say around 0.3–0.5mm (.012–.020") wide (see **Sketch 16.4B**).

Having chosen the cutter to be used – end mill, slot drill, fly cutter, etc. – and taken note of its particular requirements when being used, what other things have to be considered?

## UP/DOWN MILLING

The terms up milling and down milling originate from the days when horizontal milling was much more the norm. They are also known as climb milling (down) and conventional milling (up). Reference to **Sketch 16.5** should make the terms clear; with up milling, you contact the workpiece low and finish high, and down milling is the reverse. The sketch shows that this is achieved not by reversing the cutter but by reversing the feed direction, resulting in up milling opposing the rotation of the cutter, but with down milling they are both moving in the same direction. While the terms are confusing when being applied to vertical milling, the principles about to be described are equally applicable.

The reader will I am sure appreciate that the fit of the leadscrew in the leadscrew nut, and its end bearings, will not be perfect and some axial clearance will be present, even when new. When used, the leadscrew will be pushing the nut, and therefore the table, forward, and any clearance will be in front when no cut is being taken. If down milling is being employed, it will be attempting to move the table forward to take up the clearance,

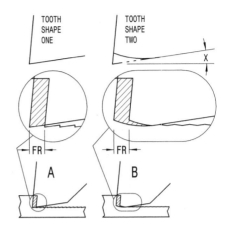

**Sketch 16.4** *Tooth shapes compared. Angle X should be only about 1–2°. FR = feed rate per tooth per rev. Hatched areas show the metal that is being removed.*

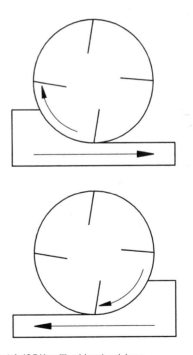

**Sketch 16.5** *Up milling (above) and down milling (below).*

and with anything other than a very light cut this will inevitably occur. The result of this is that the depth of cut increases suddenly and significantly, maybe with disastrous results. Because of this, down milling, where both cutter and workpiece are moving in the same direction, should be avoided when using typical home workshop machines. However, there is an advantage in using down milling, but unfortunately the average home workshop cannot benefit from this, except in one special situation.

**Sketch 16.6** shows, exaggerated, the difference in the cut being taken by the two methods, with up milling starting very thin (almost nothing) and finishing thick, while in the case of down milling it starts thick and finishes very thin. Even if the cutter is razor sharp, it may struggle to get under the surface in the case of up milling and one or two cutting edges may fail to make a clean cut. With further attempts as the cutter rotates and the feed is increased, a cut will be made, this intermittent situation resulting in a less than perfect finish. This is not to say that the finish is unacceptable, and in the average home workshop where the alternative, down milling, cannot be tried then the user will probably be unaware that an improvement is possible.

In the case of down milling, the cut starts thick, as the sketch shows, and therefore will more readily make the first cut, even with a less than sharp cutter. If the reader needs clarification regarding using a blunt cutter, then I feel it should be obvious that, typically, you cannot start a cut 0.001mm deep if the cutter has a 0.002mm radius on its cutting edge, as it will just skid.

Many industrial machines are backlash free and can benefit from using down milling to improve the resulting finish. There is, though, one situation where down milling can be employed in the home workshop if carried

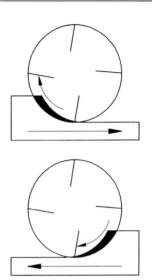

*Sketch 16.6* Comparison of cut taken.

out with care, and that is when finishing the edge of a workpiece using the side only of an end mill.

In this case, the cut can be taken using the cutting edges nearest the cutter's shank, benefiting from increased rigidity, but more importantly it will be the part of the cutter that rarely gets used so it should still be very sharp. For this method to be acceptable, the depth of cut should be small, say no more than 0.05mm (.002"), so is only usable for finishing. For it to work, the friction in the table's movement has to be enough to withstand the cutter's forces. A larger, heavier machine will be more able to benefit from machining an edge in this way, but for a smaller machine the traverse can be temporarily stiffened a little by using the table lock.

**Photograph 16.4** shows two parts, one (left) that I have machined using down milling and the other (right) using up milling. The photograph should exhibit, printing limitations permitting, a distinct difference between the two, with that on the left being distinctly

**Photograph 16.4** *Two pieces of steel, one having been finished using down milling (left) and the other using up milling (right). The one on the left has a superior finish, but down milling should be limited to very shallow cuts.*

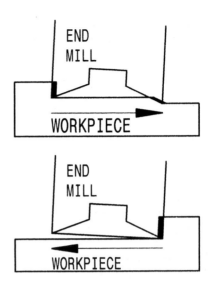

**Sketch 16.7** *Effect of traverse direction.*

smoother. While there would appear some difference along their lengths, this is due to the angle of the lighting – in both cases the finish is consistent end to end. However, this factor of the lighting shows up two aspects of the finish in the case of up milling. In the foreground the roughness is evident, but at about halfway along the rippled nature of the finish is also evident. If the edge of a fingernail is run along the length of the part, then up milling feels distinctly corrugated while down milling feels almost glass smooth. The parts are 12mm (approximately ½") wide, the cutter 20mm (¾") diameter and its speed 400rpm, the depth of cut being 0.2mm (.008"), and both were done with the same cutter.

In practice, the benefit can often be achieved by returning the workpiece with the machine still running but without increasing the depth of cut. In this case, the minute deflection of the tool when making the previous cut will provide just sufficient for the effect to be realized, even though it will be much less than the 0.2mm I referred to above.

The reader may question making a 12mm wide cut on a relatively light duty machine, but due to the cutter's helix and the shallow cut the width of the cut is considerably less. Note that the improved finish achieved only applies to faces machined using the side of the cutter.

## CUTTER TILT

As machines, even precision ones, are made to permitted tolerances, it is almost inevitable that the machine spindle is not truly perpendicular to the machine's table traverse, and because of this one side of the cutter will be lower than the other. As a result, there will be marginal differences in the way the cutter performs depending on the traverse direction. This can be seen in **Sketch 16.7** where in one direction it cuts just on its leading edge, while in the other the trailing edge will also make a cut; this is called back cutting. **Photograph 16.5** (right) shows an example of this, with that on the left having been carried out in the reverse direction, that is with the trailing edge high, and having a much superior finish. The test was carried out using a new end mill.

The photograph does, though, show that even in the reverse direction there is

evidence of both edges leaving a mark on the workpiece, which is theoretically impossible but is no doubt due to vibration causing the head to bounce up and down minutely. Even so, the finish is far superior. No doubt with the more robust industrial machines this effect would be much less likely.

The difference in height between the leading and trailing edges will be very small and therefore the trailing edge will be called upon to make a very shallow cut, and as I explained above a cutter will struggle to make such a cut. The effect of back cutting is therefore much greater when a less than new cutter is used. This can be seen in **Photograph 16.6** using a slightly blunt cutter where the same two tests have been carried out. This does show that even though that on the left still has a superior finish there is more evidence of it cutting in both directions. I suspect that the reason for this is twofold: the blunter cutter will result in more vibration, but probably more significant is that the leading edge will produce more of a furrowed result, with the crests being higher than they would be with a sharp cutter. In this case, the trailing edge would be more likely to reach the surface.

*Photographs 16.5 (above) and 16.6 (below) Test being carried out in both directions with new (16.5) and slightly blunt (16.6) cutters. The results show that the finish depends on the feed direction, that on the left being the best in both cases.*

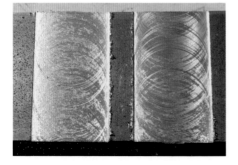

I mentioned above that the difference in the height of the leading and trailing edges would be very small. To quantify this, therefore, in Chapter 7 I explained that the specifications state a maximum error in the vertical axis of the machine spindle should be no more than 0.025mm over 300mm (.001" over 12"). If the error was at this maximum level, the difference in height of the two edges for a 20mm (¾") cutter would still only be 0.0017mm (.00007").

The explanation shows that where finish is important there is likely to be one direction in which the finish is better than the other. However, as international standards for machine tools permit the angular error to be either way, you will have to run a few test cuts to determine which direction this is for your machine, or set up a DTI to test the situation as was shown in **Photograph 7.3**.

If, however, you have a machine which permits you to set the angle, then see Chapter 7 for more detail. In this situation, my preference would be when feeding the table right to left (increasing numbers on the leadscrew dial on most machines) for the left-hand side of the cutter to be the highest, therefore minimizing back cutting in that direction.

There will, of course, be instances where the demands of the workpiece do not permit, or do not benefit from, all machining to take place in the same direction. Most important, though, will be when a large area, that is visible in the final assembly, is being surfaced, and in this case making each pass in the direction that minimizes back cutting is worth attempting. Do remember that a sharp cutter will improve surface finish considerably, and it is worth keeping a cutter specifically for the task of finishing a surface. Chapter 17 contains a simple device that will enable the end cutting edges of an end mill, that most affect surface finish, to be sharpened very easily.

The above explanations have all related to the workpiece moving left and right, but will of course apply to it moving towards and away as errors in the spindle angle can also occur in this direction.

## HOLLOW SURFACES

Another effect of the spindle not being perpendicular is that the cutter will cut a shallow trench, this being deeper the greater the diameter of the cutter being used. The effect will be minimal but is worth being aware of, especially when using a large cutter, typically when fly cutting. The effect becomes more important when machining large

surfaces that have then to be scraped flat, a surface plate, or a machine slide. In this case, multiple passes with a smaller cutter will minimize the amount of scraping necessary.

A different effect occurs if the spindle error is across the line of the traverse; in this case, the resulting effect takes the form of a shallow saw tooth. Again the effect is minimal, but this is a situation that I am more aware of in my workshop. This will also be more pronounced with a larger diameter cutter.

## SPEED

Having set up the workpiece, decided on and fitted the cutter to be used, and chosen from which direction the workpiece will approach the cutter, it is still necessary to decide the speed to be used. This, in theory, is dependent on a number of factors, although some are only of importance in an industrial situation; for completeness, however, I will mention them all.

**(1)** A higher speed will reduce the time taken to make the component.

**(2)** A higher speed will reduce the working life of the cutter.

**(3)** The additional time taken due to additional tool changes may be greater than the time saved by using a higher speed.

**(4)** Is a flow of coolant to be applied to the workpiece and cutter?

**(5)** Choice of speed will depend on the surface finish required.

**(6)** Choice of speed will depend on the machine being used – is it robust or otherwise?

**(7)** Sharper cutters will in most cases permit higher speeds to be used.

**(8)** The material being used for the part will affect the speed that is suitable. Typically, steel type 230M07Pb (leaded) will machine easier than 230M07 (non-leaded).

In the home workshop only (5) to (8) are of real importance, with (2) being a minor consideration (but only if the cutter is run well over its optimum speed). From this, therefore, the reader will understand that no list of speeds can be better than a very rough guide, as the compiler will not know the robustness of your machine, the surface finish required, how sharp the cutter being used is, etc.

For the home workshop I will therefore suggest the following as a basis on where to start when using an HSS end mill. With experience, you will certainly find that there is a wider margin for choice of speed than many charts would have you believe. This becomes obvious when you consider that in the past fine work was carried out on lathes that had just three speeds, or the luxury of six if fitted with a set of back gears.

My starting point is, therefore, when using a sharp 12mm (or ½") end mill on mild steel (230M07), use a speed of 500rpm or the nearest lower speed available. From here, attempt to keep the peripheral speed constant, that is typically 1000rpm for a 6mm cutter and 250rpm for a 24mm cutter. With experience of the machine being used, you will soon become more confident as to the speed to be used. With leaded steel (230M07Pb), you will find that a higher speed will be acceptable, say +20%.

For cast iron and bronze, halve the speeds, and for aluminum they can be doubled. For blunt cutters I would suggest reducing to 75% of the speed I am suggesting, although using blunt cutters on aluminum is best avoided, as aluminum requires a sharp edge and a generous rake angle.

If you are using carbide tipped cutters, then the above speeds can in theory be doubled, at least, and a generous feed rate applied, but this will of course depend on whether your machine has these higher speeds available and can also cope with the increased load. Only the user of a particular machine can determine this.

## FEED RATE

As very few workshops will have a calibrated power feed, my own included, quoting metres per minute would be pointless. I would suggest, therefore, as nothing will be lost other than in the time taken, that you feed the workpiece at a slow rate – if too fast, then either the machine will complain or the finish will be poor. If both are OK, try increasing the feed rate a little and observe the results. You will soon learn what is and what is not acceptable.

If when machining steel the swarf produced has a bright blue appearance, the speed and/or the feed rate is too high, and your high speed steel end mill will soon lose its cutting edge. If, however, you are using carbide tipped tools to their limit, this is much less of a problem; in fact it is expected, as they can withstand the high temperatures being developed.

Finally, we get to the point of making the first cut. Do, though, make a final check on the security of the workpiece and that all the feeds not being used are firmly locked. This includes the head on its round column, which from experience I find it easy to overlook. With that done, you can press the button and at last start removing metal.

CHAPTER 17

# MAINTAINING YOUR CUTTERS

Sharpening milling cutters varies from being practical in some cases and totally impracticable in others. Note that I do not say impossible – it is just that the effort involved is often not warranted by the costs saved.

Having been guilty myself of carrying on machining when a cutter becomes blunt, it is probable that others have done this also. From experience, though, I have learned that using sharp tools is much more rewarding and produces better results.

### END MILLS

One area that is practical is to sharpen the end teeth of an end mill once a simple jig has been made for the task, and as these are the ones that predominantly determine the surface finish achieved it is well worth doing. **Photograph 17.1** shows a possible attachment for the bench grinder that will easily perform the task. In simple terms, the cutter is placed into a square holder which is used to index the cutter for each edge by just turning it onto the next face. The cutter is then plunged into the front face of the wheel, the depth being limited by an adjustable stop, the round item to the left of the cutter in the photograph.

*Photographs 17.1 A fixture for sharpening the end cutting edges of an end mill.*

The reader may question this method of sharpening the edges on the basis that the front face of the wheel may not be that flat. However, the cutter only cuts on the extreme tip by an amount equal to the feed rate per tooth, being no more than 0.05mm (.002") even for a roughing cut. Because of this, the state of the edge beyond that is of limited importance. The only important requirement is that the end of the cutter is concave (C) by about 1–2° (see **Sketch 17.1**).

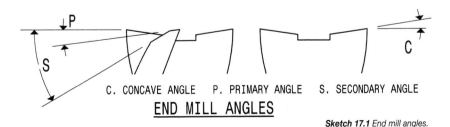

C. CONCAVE ANGLE    P. PRIMARY ANGLE    S. SECONDARY ANGLE

## END MILL ANGLES

*Sketch 17.1 End mill angles.*

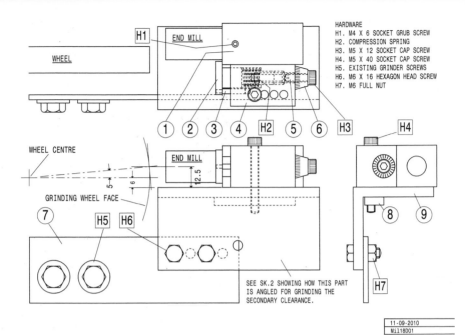

HARDWARE
H1. M4 X 6 SOCKET GRUB SCREW
H2. COMPRESSION SPRING
H3. M5 X 12 SOCKET CAP SCREW
H4. M5 X 40 SOCKET CAP SCREW
H5. EXISTING GRINDER SCREWS
H6. M6 X 16 HEXAGON HEAD SCREW
H7. M6 FULL NUT

WHEEL

END MILL

H1

WHEEL CENTRE

END MILL

GRINDING WHEEL FACE

SEE SK.2 SHOWING HOW THIS PART
IS ANGLED FOR GRINDING THE
SECONDARY CLEARANCE.

11-09-2010
Mil18D01

*Sketch 17.1* Sharpening end mills.

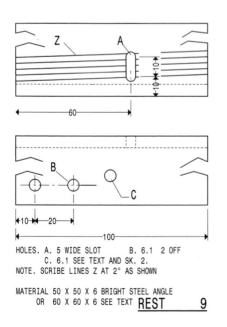

HOLES. A. 5 WIDE SLOT    B. 6.1  2 OFF
        C. 6.1 SEE TEXT AND SK. 2.
NOTE. SCRIBE LINES Z AT 2° AS SHOWN

MATERIAL 50 X 50 X 6 BRIGHT STEEL ANGLE
    OR  60 X 60 X 6 SEE TEXT  **REST**      9

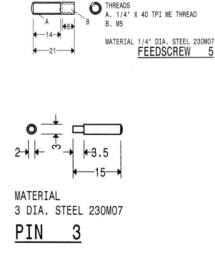

THREADS
A. 1/4" X 40 TPI ME THREAD
B. M5

MATERIAL 1/4" DIA. STEEL 230M07
**FEEDSCREW    5**

MATERIAL
3 DIA. STEEL 230M07

**PIN    3**

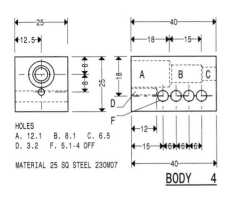

HOLES
A. 12.1    B. 8.1    C. 6.5
D. 3.2    F. 5.1-4 OFF

MATERIAL 25 SQ STEEL 230M07

BODY    4

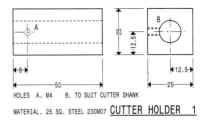

HOLES  A. M4    B. TO SUIT CUTTER SHANK

MATERIAL. 25 SQ. STEEL 230M07  CUTTER HOLDER    1

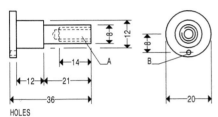

HOLES
A. 1/4" X 40 TPI ME THREAD
B. 2  SLIGHTLY COUNTER SUNK ON OUTER SIDE

MATERIAL. 20 DIA. STEEL 230M07    STOP    2

MATERIAL 12 X 6 STEEL 070M20    NUT    8

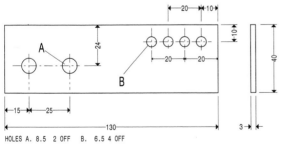

HOLES A. 8.5  2 OFF    B.  6.5 4 OFF

MATERIAL 40 X 3 STEEL 070M20, OR SHEET STEEL

NOTE. THE DIMENSIONS GIVEN SUIT A BLACK AND DECKER PROFESIONAL GRINDER
AND MUST BE CHECKED FOR ANY OTHER MAKE.    SUPPORT    7

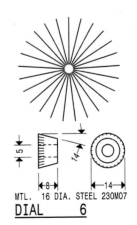

MTL.  16 DIA. STEEL 230M07
DIAL    6

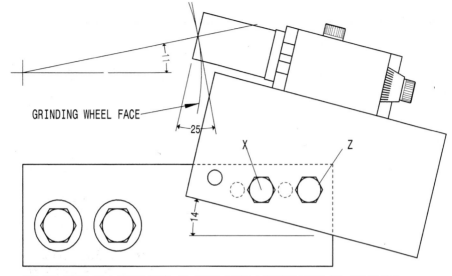

GRINDING WHEEL FACE

THE 14° ANGLE IS A GUIDE ONLY, DO MOUNT USING A SINGLE SCREW "X" AND ADJUST
TO ACHIEVE THE 25° AS SHOWN. THEN MARK POSITION OF HOLE IN THE TABLE USING
THE HOLE ALREADY IN THE SUPPORT AND DRILL FOR SECOND SCREW Z.

**Sketch 17.2** *Setting the secondary clearance angle.*

**Photograph 17.2** *The adjustable stop on the left sets the dept of the cut.*

## IN USE

Fix the cutter into the cutter holder with its cutting edges horizontal and vertical, and place it on the table such that the right-hand edge of the wheel just clears the cutter's lower vertical edge. At the same time, set the body at an angle of 2° by placing it parallel to one of the lines on the table and secure in place (see **Photograph 17.2**). The stop is clearly seen in the photograph, and the lines on the table should also be visible.

Set the stop so that the cutter is just clear of the wheel, and with the wheel then running advance the cutter using the stop until the cutter just touches the wheel. Remove the cutter and holder and advance the stop by 0.025/0.05mm (1–2 marks on the dial) and plunge each edge in turn at that setting. Repeat as necessary until a sharp edge is achieved. This should take no more than five minutes once the fixture has been made.

If the front of the wheel is in a poor state, then dressing it prior to using it for the task should be done. In doing this, ensure that the face of the wheel is parallel with the front edge of the table, making sure as a result that the 1–2° angle is achieved.

Eventually, the primary clearance (P) face will become too wide and the secondary clearance (S) will require grinding. For this, the process is the same except that the table has to be tilted. Neither the primary clearance (5°) or the secondary clearance (25°) angles are vital but do not increase the 5° value appreciably as this will produce a finer edge that will more readily become blunt. To check the set-up for these, cut pieces of card with internal angles of 95° and 115° and mark them at 12.5mm above their base. If the base is stood on the table, the sloping edge should contact the wheel close to the mark.

### SETTING THE ANGLES

The main assembly drawing shows how, working to the 6mm dimension the 5° angle is automatically achieved – again I stress that the angle is not critical! **Sketch 17.2** shows how the 25° angle is achieved by repositioning the table.

### SLOT DRILLS

As one cutting edge is longer than the other, the method for sharpening slot drills has to be modified slightly. Firstly, set the position of the cutter relative to the wheel using the shorter edge and grind both edges as was done when sharpening an end mill. This, however, results in the longer edge only being sharpened over part of its length. To fully sharpen this edge, set the guide to an increased angle of about 4° and just sharpen its center portion, ensuring that you do not remove any from the edge's tip as it will no longer be at the same level as the shorter one.

When being traversed, a slot mill works much like an end mill except that it has just two cutting edges and needs to be traversed rather more slowly. However, while being plunged, a slot drill cuts over the full length of the end cutting edges, so the condition of these edges is important.

### MANUFACTURE

This is straightforward except that the support (7) will almost certainly need some dimensional changes to suit the grinder on which the attachment is to be mounted, as this is using the fixings for the grinder's normal rest. All other parts remain as drawn.

**Photograph 17.3** shows a method for machining the ends of the table, but it may be considered adequate to just leave these as a sawn edge as a time saver. I have included the picture as it shows yet another example of using the angle plate and a cylindrical square.

Calibrating the dial can be done using some form of dividing device, but if this is not

*Photograph 17.3 Machining the ends of the rest (9). Some may choose to leave these in the sawn state, as a time saver.*

available, or the reader wishes to complete the task quickly, the following will produce an acceptable result. Mark just the edge of the dial with marking blue, say 2mm wide, and place it centrally over the template provided with the drawings and scribe a line at each point. This is maybe not very elegant, but more than adequate for the task.

If you wish to simplify the manufacture still further, you may like to consider mounting the rest using just two pieces of timber (see **Photograph 17.4**).

### SLITTING SAWS

These can be sharpened, but it is obviously not a free hand task, and needs some simple equipment to achieve it. It is more practical to do on large tooth saws, but is not impossible on smaller ones. **Photograph 17.5** shows a larger tooth saw being sharpened. With this the saw is indexed using a leaf spring and, together with the base on which it is mounted, moved left and right while being held against a fence. The fence is largely below the saw but can just be seen above and to the right of it. Should you like more details of this and sharpening other workshop tools, 'Tool and Cutter Sharpening', Workshop Practice Series number 38, may be of use. As well as guidance into the methods of sharpening tools, it also has the design for a useful grinding rest that will produce results equal to a tool and cutter grinder in many cases. With a few simple additional items, though, the set-up seen in the photograph could easily be carried out on the rest detailed in this chapter.

Photograph 17.4 Another time saver would be to mount the fixture off a simple timber structure rather than the support (7).

Photograph 17.5 Sharpening a slitting saw.

# CHAPTER 18
# FINAL COMMENTS

Having now arrived at the final chapter of the book I hope the reader has followed the trend that I was attempting to convey regarding using the home workshop milling machine. That is, after equipping the machine the process of using it for anything other than the simplest task is, typically, 40% choosing the method, 40% setting it up, and only 20% actually machining the part. Choosing the method being a major part of a task is largely due to the very wide range of workpiece shapes likely to be confronted. Because of this, mastering this particular aspect of using a milling machine is by far the major part of learning to use the machine and no book, this one included, can complete this task. I do, though, hope the examples that I have used to illustrate this feature will enable the reader to make some progress in this direction.

One modern feature of using a workshop machine is the application of CNC techniques, and the reader will have found that this has not been given an airing in the book. My reason for this is that I consider it is a facility to be added to an already established workshop by the addition of a CNC enabled machine rather than something to equip the workshop with from its inception, or even to be acquired as a replacement to a manual machine. I do believe that the workshop owner needs to have a hands-on knowledge of using a machine before entering this field – some of course may not agree. However, to have added in-depth details regarding its use would have expanded the book beyond what would be acceptable for the size of the book, and beyond what I consider the majority of readers would require. The reader with an interest in the subject should therefore seek further reading.

Having attempted to present the theory, then if you would like some hands-on experience 'Milling, A Complete Course', Workshop Practice Series number 35, could be a starting point.

# INDEX

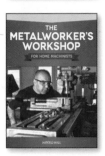